IPM Implementation Workshop for West Africa

Workshop Proceedings

Accra, Ghana, 1992

Published for the
Integrated Pest Management Working Group
by the
Natural Resources Institute
the executive agency of the Overseas Development Administration

Cover photograph
Farmers prioritizing farming problems/constraints in Upper West Region, Ghana, May 1993
Hilary Warburton

Illustration
Neoaplectana carpocapsae (Steinernematidae)

No charge is made for single copies of this publication sent to governmental and educational establishments, research institutions and non-profit-making organizations working in countries eligible for British Government Aid. Free copies cannot normally be addressed to individuals by name but only under their official titles. When ordering please quote **PSTC8**.

Natural Resources Institute
ISBN 0 85954–378–1

ii

Contents

SUMMARY

A workshop organized by the Integrated Pest Management Working Group (IPMWG) was held in Accra from 27 April to 1 May 1992. The aim of the workshop was to discuss the implementation of IPM in West Africa, the constraints to its implementation, and to put forward initiatives that could be implemented at the regional, sub-regional and national levels. The workshop was attended by regional representatives, national and international research and extension officers and representatives from NGOs and donor agencies.

Representatives from the Sudano-Sahelian, Humid and Sub-Humid West African sub-regions identified the crop/cropping system and pest priorities in their sub-region, and considered the constraints to the implementation of IPM and initiatives for overcoming these constraints. Political, institutional, socio-economic and technical constraints were identified by each group.

The main political constraints identified included:
(a) lack of awareness/recognition by policy-makers;
(b) lack of communication between farmers, technicians and policy-makers;
(c) low national budgets;
(d) lack of appropriate crop protection legislation;
(e) pesticide donations and subsidies.

Initiatives to overcome these constraints included:
(a) greater efforts to make policy-makers and the general public more aware of the value of IPM;
(b) lobbying governments for crop protection legislation;
(c) more stable financial support.

At an institutional level the main constraints identified were:
(a) lack of an adequate infrastructure for research extension and information dissemination;
(b) little long-term sustainable planning;
(c) lack of co-ordination between international, regional and national institutions, including non-governmental organizations (NGOs).

Initiatives included:
(a) the creation of national IPM Working Groups to co-ordinate activities;
(b) carefully targeted training programmes.

The main socio-economic constraints were identified as:
(a) low levels of farmer involvement in the development and implementation of IPM;
(b) lack of information on the socio-economic environment of the farmer;
(c) inadequate support for IPM extension;
(d) lack of available inputs.

Initiatives to overcome these constraints included:
(a) farmer training and on-farm demonstration;
(b) greater recognition of socio-economic factors in the development of IPM packages;
(c) emphasis on low-input, cost-effective IPM strategies;
(d) the transfer of chemical subsidies for the provision of appropriate inputs.

Finally, technical constraints included:
(a) incomplete data on pests and ecosystems;
(b) lack of co-ordinated research and evaluation.

Initiatives identified were:
(a) the development of clear research plans aimed at specific problems for submission to donors;
(b) pest diagnosis and ecosystem analysis with improved training and information dissemination at all levels.

Countries provided action plans outlining priority initiatives for the next 15 years. These highlighted the importance of national governments taking the necessary initiatives in promoting IPM. It was recognized that IPM research must be better co-ordinated and information services improved. Many countries gave high priority to involving and training farmers in IPM, clarifying precise research aims for IPM studies at an early stage and enhancing knowledge of existing crop and pest situations. The action plans prepared by each country were to serve as a basis for discussion and the initiation of national IPM programmes.

The delegates agreed that the constraints common to IPM implementation showed a basic need for a network which would integrate experience and pool information. Such a network would include national services, private institutions, NGOs and international institutions and agencies. It would serve to improve awareness of IPM at the sub-regional level, facilitate the identification of research and extension needs and ensure against unnecessary duplication. Once common problems were identified, networking could be achieved through regular meetings, the establishment of a steering committee and a newsletter. It was agreed that any new network should be designed to reinforce existing networking facilities. Periodic evaluations should be included in the networking plan to provide a guarantee of credibility and efficiency for members and donors.

Conclusions and recommendations of the workshop were as follows.
- Government commitment is essential to the successful development and implementation of IPM. Steps must be taken to promote awareness of IPM within government circles.
- Given problems associated with the free distribution of pesticides, a system should be introduced whereby farmers contribute financially to the costs of pesticides and their application. In addition, a system of accountability for the distribution and use of pesticides should be introduced.
- IPM is a component of integrated crop production and should be developed and implemented within the context of agricultural production systems. More attention should be given to the growing of healthy crops as the basis of IPM.
- IPM programmes should be planned with farmer involvement, based on preparatory studies of the socio-economic context in which farmers operate, with special emphasis on gender issues.
- Systems of monitoring and cost/benefit analysis should be an integral part of IPM packages.
- The collaboration of NGOs in the promotion and implementation of IPM should be encouraged.
- Clearly defined medium- and long-term strategic plans, necessary for donor support, should be prepared at the national level.
- Regional IPM networks should be created and existing networks strengthened.

INTRODUCTION

The Integrated Pest Management Task Force was constituted in 1989 to report to the Technical Advisory Committee of the Consultative Group for International Agricultural Research (CGIAR) on the status and implementation of IPM in developing countries.

A consultants' report published in 1992 entitled *Integrated Pest Management in Development Countries: Experience and Prospects*, provided a broad overview of major issues influencing IPM in the developing world and made numerous recommendations for improving IPM implementation. The increasing focus on initiatives for implementation has resulted in the reconstitution of the Task Force as the Integrated Pest Management Working Group (IPMWG). This group has organized a series of regional workshops involving key developing country researchers, extension workers and policy-makers.

This report contains the proceedings of one of these workshops organized to validate the conclusions of the consultants' report through discussion with IPM specialists and policy-makers in West Africa. The workshop was held in Accra from 27 April to 1 May 1992. It was attended by national representatives and delegates from national and international research and extension institutions, non-governmental organizations (NGOs) and donor agencies. A complete list of participants and IPMWG members is included at the end of the report.

OBJECTIVES AND OUTPUTS OF THE WORKSHOP

The objectives of the workshop were:
(a) to determine the relevance to West Africa of the synopsis of the IPMWG consultants' report through discussion with workshop participants;
(b) to identify the current constraints to the development and implementation of IPM strategies in West African countries;
(c) to propose initiatives needed to strengthen national capabilities in IPM development and implementation.

The workshop participants were divided by country into two sub-regions and encouraged to prepare proposals for Country Action Plans for consideration by donors. This resulted in the following five points.
(a) A list of crop and pest priorities for IPM.
(b) A list of major constraints to IPM development and implementation.
(c) A list of initiatives needed to address constraints; attention was given to the relative strengths of different countries in the sub-region in policy resources and experience needed to implement each initiative.
(d) Phased 15-year Country Action Plans for the implementation of the identified initiatives.
(e) Proposals for improved networking within the region.

OPENING CEREMONIES

Welcoming remarks

M.J. ILES

IPMWG Secretary, Natural Resources Institute, Central Avenue, Chatham Maritime, Kent ME4 4TB, UK

On behalf of the IPM Working Group I want to welcome you all to the workshop. In addition to IPMWG member support, the following generously supported participation at the workshop: Food and Agriculture Organization of the United Nations (FAO), Technical Centre for Agricultural and Rural Co-operation (CTA), International Development Research Centre (IDRC), United Nations Development Programme (UNDP) and United States Agency for International Development (USAID).

The large number of participants at this workshop demonstrates the importance of IPM in West Africa, and I think that the most important aspect of the workshop will be the opportunity for dialogue, both formal and informal, between the country representatives, donors, international institutions and NGOs present.

The main objective of the workshop will be to identify initiatives to improve the implementation of IPM in the region. I urge you not only to discuss the techniques and technical constraints to IPM, but to use the workshop to identify the wider social, economic, political, institutional and information exchange constraints to IPM implementation, and develop practical strategies for their removal. The presence of numerous donors at the workshop provides country representatives with an ideal opportunity to make them aware of areas in need of change or assistance.

Finally, I want to stress that the ultimate aim of the workshop will be to identify strategies to increase agricultural output in West Africa, as well as to assist those whose livelihoods depend on agricultural production. You should keep the needs and priorities of farmers themselves foremost in your minds throughout the workshop.

Objectives of the workshop

F. VICARIOT

IPMWG, Office de la Recherche Scientifique et Technique Outre-Mer (ORSTOM), 213 rue Lafayette, 75480 Paris, France

This workshop was organized by the IPMWG, the main aim being to encourage researchers within the international community to provide development agencies and agriculturalists from developing countries with the scientific knowledge necessary for sustainable agriculture. IPM plays a major role in this.

One can say that IPM deals with 'techniques which facilitate better control of parasitic activity on crops'. It is making use of available opportunities for the implementation of various aspects of the IPM strategy at the appropriate levels (i.e. crop, regional, national and sub-regional).

To achieve this, the IPMWG decided the following.

(a) To start on the basis of the experience acquired in each West African country (21 countries in three sub-regions, namely Sahelian Africa, the Humid and Sub-Humid sub-regions) by inviting, from each country, an official in charge of research and another in the field of development/extension work.

(b) To enter into dialogue with donors likely to assist in the execution of projects or action plans at levels which will have to be identified.

(c) To invite some experts on this subject with the aim of enriching our knowledge especially at the working group level.

(d) To invite the region's international centres, e.g. the International Institute for Tropical Agriculture (IITA) and West African Rice Development Association (WARDA), and NGOs involved in such activities.

The objectives of this workshop are:

(a) to identify crop and cropping system priorities at the sub-regional level for the implementation of IPM programmes;

(b) to draft an action plan for each sub-region for these crops and present these plans to countries and donors;

(c) to identify constraints to the implementation of these programmes on the basis of experience acquired in various countries;

(d) to identify initiatives to be taken to eliminate these constraints at the political, institutional, socio-economic and technical levels;
scientific (multidisciplinary)
institutional : co-operation between various departments concerned
regional co-operation based on comparative advantages of international, national and regional institutions
project planning level : training, programme formulation
attempt to identify with respect to each sub-region and some crops, the basis of an action plan which would be presented to countries and donors—these action plans would be drawn up on a medium-term basis (10–15 years)
explore avenues for networking at the sub-regional level at least with regard to certain aspects of the action plan.

To achieve these objectives, the workshop is organized into two sections. The first two days will be devoted to country papers in which speakers will describe their experiences of IPM in their own countries. In the second part working groups made up of participants from the different sub-regions will discuss the points arising from the first part of the workshop.

The first IPM Workshop was held in September 1991 in Southeast Asia and was a great success. Another workshop will be held in early next year in Harare, Zimbabwe and will have similar aims and objectives.

Welcoming remarks
R.T. N'DAW
Assistant Director-General, FAO Regional Representative for Africa

On behalf of the Director-General of FAO, and on my own behalf, it is my pleasure to welcome you all to the IPM workshop for West Africa, which has been organized by the IPMWG, of which FAO is an active member.

As you have already been informed by previous speakers, the IPMWG was formed to review the status of IPM implementation in developing countries. It commissioned a report on the experience and prospects for IPM in Third World agriculture by a group of consultants. The report provides a broad overview of the major issues which influence the pace of IPM adoption in developing countries, its concept, progress and constraints.

The current focus of the IPMWG is to identify those initiatives with the potential for successful implementation through a series of regional workshops such as this one. This workshop brings donors and West African institutions together to discuss the implementation of IPM. The donors have come for a specific purpose, to find out the interest at a national level for IPM and priorities for action. West African institutions concerned with plant protection have an equally well-defined reason for attending: their ability to work on plant protection in the future will depend in part upon their success in persuading donors that pest management technologies can help farmers.

The international development community has shown increasing interest in promoting an IPM approach to crop protection problems. Over-use of pesticides has become a worldwide concern and IPM is commonly accepted as essential in keeping their use in check. FAO has been active in this field since the 1960s and the World Bank has recently adopted IPM as its central strategy for crop protection projects. The climate for IPM is clearly favourable.

Although as a concept IPM has been around for nearly 30 years, its development and adoption in West Africa has been disappointing, despite its many apparent benefits. IPM is still not featuring in farmers' cropping systems.

6

This situation has been attributed to a number of constraints operating across the entire technical, socio-economic and political environment in which the pest problem operates. Experience with FAO projects on plant protection in West Africa has shown that the key constraint to IPM adoption by farmers is a general lack of practical, effective control technologies. As important as the availability of proven technology, is a delivery system which effectively brings these technologies to farmers. This requires an extension system that works closely with researchers and farmers. This is often a major constraint in West Africa, due to the general weakness of plant protection and extension services.

Any approaches or new initiatives to promote the development and adoption of IPM in West Africa must take into consideration the fact that for most cropping systems, the basic knowledge needed to develop IPM technologies may be lacking. As a first step, therefore, IPM requires a major research input at the farm level. This would, of course, require strengthening research infrastructures. In addition, extension personnel and farmers will have to be trained and the extension services strengthened.

All these are activities that are long term and require substantial inputs of resources, both human and material. These may be beyond the means of West African governments, most of whom are presently undergoing economic adjustments. It is important that the international donor community recognizes this and gives greater priority and more funds to IPM-related projects. Some form of international initiative, which will present the need for increased funding, is certainly required if IPM is to secure a greater share of donor funding. Thus, the creation of the IPMWG is welcomed. It should also be congratulated for the various activities already initiated, including this workshop.

If progress is to be made in IPM implementation, governments must act by adopting policies that involve IPM as an integral part of their agricultural development programmes. We cannot convince donors to put more funds into IPM if governments themselves do not show a strong commitment to developing and implementing IPM.

I sincerely hope that the workshop will focus attention on some of these key issues, and provide some answers to the major constraints that governments are likely to face in their attempt to develop and implement IPM policies. FAO attaches great importance to the development and adoption of IPM in Africa and will continue to assist its member governments in their efforts to develop and promote good crop protection policies.

May I take this opportunity to thank the honourable Deputy Secretary of the Provisional National Defence Council for the Western Region, Dr Francis Abu, for presiding over the opening of this workshop. Please convey our most sincere thanks to the Government of Ghana for having agreed to host this workshop. As always, we appreciate the generosity and hospitality of the Ghanaian people.

To the participants, observers and resource persons, I wish you a very successful workshop.

Official opening

J.F. ABU
Deputy Secretary for Agriculture (Western Region), Ghana

IPM is a dynamic and evolving system which supports the co-existence of pests, their natural enemies and other requisites, compatible with productivity and environmental quality, in man-made and man-managed agro-ecosystems. The concepts and practices of IPM have resulted mainly from the disillusioning experiences with pesticide use, the undesirable side-effects of which have provoked various reactions. Some environmentalists call for a total ban of pesticides, but it is strongly believed that pesticides will continue to play a key role in pest management. The challenge we face is to be able to increase agricultural production without destroying the environment.

Pest management fundamentally recognizes the existence of certain biological forces in nature which influence and determine increases and decreases in pest populations. These forces can be manipulated and managed within reasonable limits to ensure that the damage and losses due to pests and other harmful organisms are minimized. Certain artificial methods or inputs may be introduced to contribute to the management of the system, provided a full appreciation of the individual and composite role of these inputs are made. Pest control should not attempt to eliminate damage or loss; its main aim should be to combine the economic and biological factors with other methods of control into a system which is as close to natural control as possible. IPM is now the basic approach to most pest situations in developed countries. In these countries the concept has developed sufficiently to give rise to very precise economic thresholds and injury levels, and sophisticated integration of biological and other approaches in harmonious combinations for various pest situations of agricultural and medical importance. In Africa, particularly West Africa, IPM is less sophisticated, and is practised only in cropping systems that are limited to some cereals, such as rice and maize, grain legumes, particularly cowpeas, cotton and cocoa.

Although there seems to be poor uptake of IPM in Africa, peasant farmers have, consciously or unconsciously, incorporated pest management practices into their production systems for a long time. These consist of the traditional tolerance for some level of damage to the crops and the diversity of the agro-ecosystems, such as mixed cropping, methods of clearing, land preparation, taungya systems, shifting cultivation, varieties and cultivars that performed satisfactorily under moderate infestations and management systems such as the destruction of crops residues. With proper input of pesticides, where and when necessary, the traditional African farming approach could offer an excellent basis for the development of an appropriate IPM system.

It is accepted that agriculture will continue to be the main vehicle for improving the lot of peasant farmers in developing countries. Farming will remain the key factor for our economic growth and survival. This situation is expected to put much stress on our land resources for increased production. An obvious example is Ghana where the current population of about 15 million people depend on 1 million ha of land available for cultivation. By the year 2000 the same available land, assuming there is little degradation, will support an estimated population of about 20 million people. Consequently, there has to be higher production per unit area. Unfortunately the traditional agricultural systems are not readily adaptable to intensive production and radical measures will have to be adopted. Experience indicates that the technological packages available to farmers may lead to more disastrous pest problems. The implications for pest management practitioners on the African continent are obvious. It is alarming to learn that up to about 25 million agricultural workers in developing countries suffer from pesticide poisoning, of whom some 20 000 people die each year. Although many of these casualties may be due to poor handling of pesticides and lack of protective clothing, such revelations will increase the pressure for safer pest management alternatives. Our pest management scientists will have to adapt the rich experiences from the developed countries to fit into current traditional peasant farming systems, and the transformation they are expected to undergo in the future. The time is ripe for our extension services and scientists to unite to pass on the immense knowledge acquired on IPM from laboratories, greenhouses and experimental stations to the ultimate beneficiaries—the farmers.

Although IPM is built on knowledge accumulated over time, it is not necessary to wait for the integration into complete systems before implementation. For Ghanaian scientists, the Ministry of Agriculture through the crop services and the agricultural extension services departments, now have facilities for on-farm research and adaptive programmes. Our scientists should avail themselves of these facilities. These arrangements will help shorten the usual long period between research and the implementation of IPM programmes. It will show scientists the constraints under which farmers operate, and help the farmers to appreciate, identify and even manage simple pest problems in the field. With the formal inauguration of the National Agricultural Research Committee in Ghana a few months ago, we hope there will be more interest in IPM programmes.

In conclusion, I wish to say that the complexity of the environment necessitates prolonged and expensive research before an appropriate IPM programme can be achieved. It is sad to note that most developing countries may not have adequate financial resources to provide the required facilities. African governments should appreciate the fact that the prolonged dependence on pesticides by developed countries has cost the world dearly. In many instances the damage resulting from the side-effects of pesticides are either irreparable or can only be repaired at very high cost. It will be disastrous for our governments to fail to finance the necessary research for IPM systems today only to be faced with huge bills to save the environment from side-effects of pesticides tomorrow.

I want to charge African scientists to take up these issues seriously with their respective governments and press for full attention and support for IPM.

REVIEW OF CONSULTANTS' REPORT

Integrated pest management: a review of the consultants' report and its relevance to West Africa

S.S. M'BOOB

FAO Regional Office for Africa, P.O. Box 1628, Accra, Ghana

INTRODUCTION

The consultants' report commissioned by the IPMWG on the implementation of IPM activities provides a broad overview of the major issues which influence the pace of IPM adoption in developing countries—its concept, progress and constraints.

The report is reviewed within the context of IPM development and implementation in West Africa, giving due consideration to relevant issues:
- the concept and definition of IPM
- progress in IPM implementation
- constraints
- new initiatives
- role of donors and international development agencies.

DEFINITION AND RELEVANCE OF IPM

The definition of IPM has now become a subject of discussion at international meetings. As it is used in numerous different ways, there are many misconceptions about IPM. It should be noted that the concept evolved as a reaction to over-reliance on chemical pest control. Extensive use of pesticides led to the development of resistance in many key pests and increasing costs will ultimately make crop production uneconomical. IPM is increasingly finding favour as the most appropriate control strategy for use in plant protection. It is often specifically recommended for developing countries because of the potential savings in foreign currency, environmental safety and low costs to the farmer.

The term IPM is used loosely and may refer to a concept, strategy, system, method or package of pest control measures. It is also used to give an aura of credibility to recommended pest control methods or project proposals. At an operational level, IPM has remained elusive and in order to avoid confusion and ensure common understanding and agreement on the meaning of IPM, an analysis of its characteristics will be useful.

Although definitions abound, the core of IPM is the development of a set of practices that maintains pest populations at levels below which economically significant losses are caused. It emphasizes minimal intervention, particularly with synthetic biocides, and husbandry of natural regulatory mechanisms, either biological or cultural. This is the key to the difference between IPM and classical pest control practices: since 'management' implies the continued existence of the pest within a balanced system that itself imposes control, whereas 'control' suggests direct intervention with little concern for sustainability.

The IPM approach gained popularity in Southeast Asia and Central America because excessive pesticide use in these zones had led farmers into economic difficulties through pest resistance and the elimination of natural predators.

In West Africa, the majority of small-scale farmers who produce most of the food consumed there, have, for economic and other reasons, not been large-scale users of chemical pesticides. The tendency now, however, is for increased use and greater reliance on pesticides at all levels of crop production. This situation has been actively encouraged by both the governments and international aid agencies through generous government subsidies and outright pesticide donations. It fails to take into consideration the low market value usually associated with subsistence crops as compared with the high cost of pesticide application.

PROGRESS IN IPM IMPLEMENTATION IN WEST AFRICA

From a review of the country reports submitted by workshop participants, the following general observations about the current state of IPM in the sub-region can be made.

(a) Pesticides overshadow all other current technologies as a measure for crop protection.

(b) Many potential IPM strategies have been identified but very few have been tested for their appropriateness to farmers' needs.

(c) Some West African plant protection institutions have acquired experience in classical biological control, others have acquired experience in the management of pesticide campaigns; few have experience, however, of relating new pest control techniques to farmers' needs.

(d) The Africa-Wide Biological Control Programme co-ordinated by IITA and supported by a consortium of donors, including FAO, has had tremendous success in introducing successful biological control programmes in some countries as well as building up national capabilities for biocontrol. The major difficulty facing the programme at present is that national activities initiated under the programme are only sustained as long as there is external donor support.

These observations raise a number of important questions about the direction of IPM in West Africa.

Unlike Southeast Asia, where IPM is increasingly recognized as official government policy, not a single West African country has formulated a clear policy regarding IPM. This lack of clear government policy on, and commitment to IPM constitutes a major set-back to its development and implementation.

PROSPECTS FOR IPM IN WEST AFRICA

Despite its many apparent benefits, IPM is still lacking in the majority of small-scale farmer cropping systems. It can be said that traditional farming systems maintain low pest levels and therefore practise IPM. In many cases, however, these systems are under pressure and yields are threatened at a time when the demand for food is rising. If food production is to increase in Africa methods will need to change, and this will require the adoption of IPM-type crop protection techniques.

The country papers submitted by workshop participants mentioned many possibilities for pest control that had been described as promising by researchers without being evaluated for their effectiveness or appropriateness to farmers' needs. A frequent comment in the country papers was that these 'techniques-in-waiting' could be brought to implementation if more funds were available for research, pilot programmes and extension.

One probable major reason why countries have difficulty in attracting donor funds is the lack of well-planned programmes with clear objectives for research and development of IPM. Most countries have not developed strategies for IPM development and implementation.

The success of the FAO rice project in Southeast Asia is largely attributable to farmer education and a sense of their ownership and control over the new IPM technologies. This important concept of farmer ownership has yet to gain a place in IPM terminology in West Africa.

CONSTRAINTS

Constraints that limit IPM implementation operate across the entire technical, socio-economic and political environment in which the pest problem is experienced.

Knowledge base

Inadequate basic and adaptive research into IPM methodologies and poor dissemination of research results are the two most commonly experienced constraints to IPM development and adoption. Problem identification for adaptive research is often poor with little or no farmer participation.

One of the key constraints is a general lack of practical proven effective control technologies. The now defunct project on research and development of IPM of basic food crops in the Sahel, funded by the United States Agency for International Development (USAID), is sometimes cited as an example of an ambitious programme which failed due to lack of technology. The initial years of the project were spent in building up the necessary national research infrastructure and training scientists. Hence, little time was left for the development of IPM technologies. As a result there was little appropriate technology to draw upon for the extension phase of the project.

Equally important is a delivery system which can effectively bring proven technologies to farmers. This requires an extension system that works closely with researchers and farmers. Often, this is a major constraint due to the general weakness of plant protection and extension services.

Socio-economics

Subsistence farmers in West Africa are generally poorly educated in the Western sense, have limited resources and are ill served by overstretched and weak extension services. Unless IPM technologies are simple and easy to disseminate and require minimal or no extra resources they stand little chance of being adopted.

Institutional

IPM is inherently an inter-disciplinary, multifunctional approach to solving pest problems. Current institutional structures in developing countries do little to simplify the task of the farmer practitioner. The management of research, extension and technical support services are frequently operated independently of one another, in different institutions, often with conflicting goals and interests. In addition these activities almost always have inadequate resources.

Policy

An appropriate environmental policy is essential to the establishment of IPM. The need to generate self-sufficiency in food production is a powerful imperative for most, if not all developing countries, and there remains a strong tendency to support simple short-term approaches with rapid returns, such as those dependent on pesticides. The situation is complicated further by the vested interests of the pesticide industry and its undoubted political and commercial influence in developing countries. Subsidies, in particular those involving free distribution, have done a great deal to undermine the rational and judicious use of pesticides, to the producers' disadvantage.

CRITERIA FOR SUCCESSFUL IPM

The success of IPM may be judged against a number of individual criteria. The consultants' report has identified several, ranging from an economic and safe reduction in losses due to the pest, to specific criteria associated with technological developments (robustness, affordability, responsiveness to pest situation, sensitivity to the pest complex etc.).

The ultimate goal of any IPM programme is sustainable, cost-effective crop protection which is within the capabilities of users and does not harm people or the environment. For the majority of subsistence farmers in West Africa, the first goal must be to increase the level and reliability of production. IPM should be introduced if crop losses to pests present a significant production constraint.

Given the inadequate education of farmers and extension services, technologies should require little or no cash expenditure and should be easily disseminated. These should preferably be based on resistant crop varieties, improved cultural practices and biological control. Pesticide use is not excluded but should be restricted to emergency situations, and should be accompanied by measures to prevent health and environmental hazards and the destruction of beneficial insects.

Role of donors

The international development community has shown increasing interest in promoting an IPM approach to crop protection problems. Pesticides are a major concern to environmentalists and IPM is commonly accepted as a satisfactory way of controlling their use.

The workshop may wish to give due consideration to the set of guidelines established by the thirteenth session of the FAO/United Nations Environment Programme (UNEP) panel of experts on IPM. The guidelines are for defining priorities and in making the decisions as to which crop and in which crop situation, or in which region IPM projects are most needed.

The guidelines, among other things, recognize the need to concentrate efforts on crop situations in which there is over-use or abuse of pesticides and for which an adequate knowledge base or local research support is available to help step-by-step improvement of the IPM technical input.

A careful look at the history of donor-funded IPM projects in West Africa, such as the IPM of food crop pests in the Sahel, should provide a good basis for discussions.

CONCLUSIONS

The development and implementation of IPM is of primary importance in achieving sustainable crop production in West African countries. Despite the many apparent benefits of IPM and the potential technologies that have been identified ('techniques-in-waiting'), there has been little appreciable progress in the adoption of IPM in West Africa.

Most West African countries lack the required plant protection research, extension and training infrastructures to carry out adequate IPM work. An important aspect therefore, is the need to strengthen the capacity of countries in these areas. Initiatives which will present the need for increased funding for IPM more effectively, and secure a greater share of donor funds, are required.

West African governments need to show a clear commitment to, and support for IPM development and implementation and adopt it as the national policy for crop protection.

Support from the international donor community is crucial in providing the necessary inputs for the development of IPM technologies, and in strengthening national capabilities for IPM development and implementation.

IPM EXPERIENCES IN WEST AFRICA

Training in crop protection: the DFPV experience

S.B. SAGNIA
Departement de Formation en Protection des Végétaux, Niamey, Niger

INTRODUCTION

The Permanent Interstate Committee for Drought Control in the Sahel—Comité Permanent Inter-États de Lutte contre la Sécheresse dans le Sahel (CILSS), was created in 1973 in the wake of the recurrent droughts of the late 1960s and early 1970s. Member states now include Burkina Faso, Cape Verde, The Gambia, Guinea-Bissau, Mali, Mauritania, Niger, Senegal and Chad.

The Département de Formation en Protection des Végétaux (DFPV) (Crop Protection Training Department) forms part of a larger venture by CILSS member states to strengthen their national capabilities in plant protection, particularly in the control of migrant and other pests of economic importance that severely limit food production in the region. The crop protection training component of this larger venture, to be executed by the DFPV, was inaugurated in 1981, and has since been carried out with financial aid and technical assistance provided by the Dutch government. This training project is particularly appropriate as an estimated 20–30% of cereal production in the Sahel (representing 1.18–1.77 million t) is lost to pests annually (Anon., 1992).

The main objective of the DFPV is to contribute to food self-sufficiency in the CILSS member countries through the training of middle-level technicians in crop protection.

The specific objectives are:

- to organize a variety of training courses in crop protection to satisfy the need for well-trained middle-level field personnel in the CILSS member countries;

- to strengthen national and regional centres of training, research and agricultural development particularly in the area of crop protection;

- to disseminate technical information in plant protection to field personnel throughout the Sahel through the distribution of brochures, technical bulletins and journal articles.

ACTIVITIES OF THE DFPV

Training of middle-level technicians

The acute shortage of adequately trained middle-level manpower in crop protection and the demand for their skills justifies this type of training as the key element in DFPV's activities. The training lasts for two years and 25 students are now admitted each year to the first year of the programme. The division of subjects between the first and second years is shown in Table 1.

Training in the subjects mentioned in Table 1 is both theoretical and practical, with more emphasis on practical aspects in the second year of the programme.

The basic strategy in the second year is to emphasize alternative pest control strategies as well as the safe, economic and efficient use of pesticides. The teaching of various components of an integrated control programme, i.e. varietal resistance, biological control, cultural practices etc., and the importance given to environmental protection through research and teaching in ecotoxicology, are indications of our intentions to promote sound pest management through training.

The training programme as a whole is adapted not only to the educational background of the students, but also to the conditions and needs of the Sahel (Sagnia, 1991). The entry level required for the course is the high school certificate or equivalent, plus two years field experience in agriculture.

At the end of the first year, students return to their own countries for nine weeks training in the field, either at an agricultural research station or at the national crop protection service. This enables students to familiarize themselves with the wide range of pest problems in their own countries and to apply the knowledge so far acquired.

Table 1. Two year training in crop protection at the DFPV

First year Subject	No. hours	Second year Subject	No. hours
Acridology I	42	Acridology II	42
English Language	56	English Language	50
Biology	28	Pesticide Application Equipment and Techniques	98
Agronomy	28	Ecotoxicology	56
Chemistry	42	Applied Entomology	126
Ecology	42	Introduction to Computer Science	56
General Entomology	168	Bird Pest Control	28
Genetics	56	Rodent Pest Control	28
Mathematics	84	Weed Science	84
Meteorology	56	Plant Nematology	70
Physics	42	Applied Phytopathology	70
General Phytopathology	140	Administrative Writing Skills	14
Phytopharmacy	70	Storage Pests	28
Technical Writing	14	Experimental Techniques (Statistics) II	42
Experimental Techniques (Statistics) I	42	Plant Virology	56
		Extension Methods	84
Total	910	Total	932

At the end of the second year, students undergo another practical training period which lasts for 15 weeks. This takes place at the DFPV under the supervision of lecturers. The work consists of research projects on a variety of themes in acridology, applied entomology, applied phytopathology, plant nematology, storage pests, rodent control, and effectiveness of insecticides and herbicides.

For the practical training in both the first and second year the students are required to submit a report which is presented and defended before a three-man panel of DFPV lecturers.

From 1984 to 1991, 125 technicians graduated through the programme (Table 2).

Training of senior technicians and agronomists

Senior technicians and agronomists are trained in a four-month specialized course in crop protection designed to upgrade the crop protection knowledge and skills of this category of personnel. The course is offered every two years and is open to employees of national crop protection services and national research institutes.

All the subjects in this course deal with applied aspects of crop protection. Subjects such as varietal resistance, diagnostics, storage and protection of seeds, plant virology, plant nematology, ecology, rodent pest control and applied entomology are optional, while others like statistics, laboratory techniques, computer science, pesticide applications, individual projects and field trips are obligatory. The system of options enables a student to choose not only subjects of personal interest, but also those that are appropriate for the duties performed in the field. An individual practical project is carried out by each student at the end of the programme.

In the two courses so far held 28 technicians have been trained.

Training of trainers

Training for trainers is a four-month programme designed for teachers that teach crop protection in agricultural schools or colleges in the CILSS member countries. The course is designed to increase the teachers' knowledge of crop protection, while improving their didactic skills through the use of teaching aids, such as overhead and slide projectors, posters, flipcharts, disease and insect collections.

Table 2. Number of graduates from the DFPV by country (1984–91)

Year	Burkina Faso	Cape Verde	The Gambia	Guinea-Bissau	Mali	Mauritania	Niger	Senegal	Chad	Total
1984	3	–	–	–	1	1	7	5	–	17
1985	4	–	–	–	1	–	5	6	3	19
1986	3	–	–	–	2	4	2	2	1	14
1987	2	1	–	–	3	–	3	4	3	15
1988	–	–	1	–	2	2	6	2	1	15
1989	2	–	–	–	2	–	5	3	5	17
1990	2	–	3	1	–	–	3	3	2	14
1991	2	–	–	1	4	–	2	3	2	14
Total	18	1	4	2	15	7	33	28	17	125

The preparation of teaching aids such as slides, work books and lecture notes by the participants for their own use is an integral part of the training programme. In addition, the DFPV provides them with the relevant didactic material prepared by its staff. This acquisition of teaching aids is of practical importance for these educational establishments where such material is inadequate or non-existent.

Twelve agricultural teachers have been trained through this programme.

Workshops

Two three-week workshops are held each year; about 20 participants are admitted to each. The theme of each workshop is decided by the directors of the national crop protection services of CILSS member countries during the annual meeting of the DFPV's Scientific and Pedagogic Committee of which they are members. All the workshops are practically orientated and deal with urgent pest problems facing agricultural production in the Sahel.

Since its inception, the DFPV has organized workshops on themes as varied as the safe use of pesticides, phytosanitary control, rodent pest control, surveillance of locusts and grasshoppers, and the ecology and identification of immature grasshoppers and egg pods. A total of eight workshops were organized between 1987 and 1991 and 173 technicians of National Crop Protection Services were trained in the different subject matters (see Table 3 for details).

Table 3. International workshops held at the DFPV (1987–91)

Year	Subject	Number of participants/Number of countries
1987	Efficient and safe use of pesticides	20/7
1987	Phytosanitary aspects of seeds	20/6
1988	Rodent pest control	20/6
1989	Phytosanitary inspection and quarantine	20/7
1990	Surveillance of locusts and grasshoppers*	24/12
1990	Bird pest control	20/7
1991	Rodent pest control	22/9
1991	Ecology and identification of immature stages and eggs pods of Sahelian locusts and grass-hoppers*	27/11

* The 1990 and 1991 workshops on locusts and grasshoppers included participants not only from most Sahelian countries, but also from neighbouring countries of Benin, Republic of Guinea (Conakry), Algeria, Tunisia and Morocco where these insects are also important pests.

Information dissemination

The DFPV's information service started operations in October 1988. It is a complementary activity aimed at meeting the documentation needs of DFPV lecturers and students and to disseminate crop protection information throughout the region. The database of documents comprises 662 monographs and 3123 journal articles. A mailing list of 1588 people forms part of the clientele.

A series of brochures, *Operational Acridology*, is distributed free to field personnel in the CILSS member countries. Users outside the region pay for these brochures at subsidized rates. Seven volumes have been published to date. The eighth volume is in preparation and will be devoted to the biological control of locusts and grasshoppers through the use of predators, parasites and pathogens.

In addition to these documents the information service prepares and distributes the following publications at regular intervals:

- a newsletter to liaise with former students, staff, workshop participants and interested plant protection practitioners;
- an accessions bulletin of titles received by the DFPV documentation centre;
- an accessions bulletin by subject matter.

Sahelian users can order free literature from the DFPV library through the accessions bulletins.

The DFPV has good up-to-date collections of workshop proceedings, together with lecture notes and laboratory manuals prepared by its own staff, together with slide collections. These are arranged according to discipline (phytopathology, plant virology, plant nematology, acridology, general and applied entomology, rodent and bird pest control etc.). There are also insect and disease specimen collections and a herbarium for teaching and reference, plus a wide variety of other teaching aids.

Research in support of training

Every training programme or institution needs research support to upgrade and update the course contents. Research activities, particularly those designed to accommodate student research projects, are valuable complementary tools to the DFPV's training activities.

A limited amount of research, with the specific objective of enriching the training programme and improving the research capabilities of the teaching staff and students, is carried out at the DFPV in the activities of the third phase of the project.

Research is currently being carried out in the following areas;

- acridology: mass rearing of several locust and grasshopper species, testing of biological control methods against locusts and grasshoppers, biomodelling and population dynamics of *Oedaleus senegalensis*;

- biocontrol of the cowpea bruchid: cowpea storage systems at village level, host range of the parasitoids, suppressive capacity of a parasitoid, and combination of traditional control methods and biocontrol.

In addition to these research activities which are already operational, similar objective-orientated activities will be conducted in other fields such as phytopathology, plant nematology, virology, applied entomology, phytopharmacy, ecotoxicology and weed science.

It is hoped that these activities will generate useful results and teaching aids, e.g. slides and subjects for laboratory or field practical lessons, that could be fed back into the training process.

A locust and grasshopper biological control project funded by the Dutch Government has been in operation at the DFPV since 1990. This is part of a larger international programme involving the International Institute of Biological Control (IIBC) and IITA. This aims at finding suitable fungal pathogens for use in the control of grasshoppers and locusts in the hope of reducing the amount of pesticides used for this purpose. Results generated by this project will be used to reinforce our training programme, particularly in the area of biological control.

CONCLUSION

The DFPV, because of its unique character, plays a major role in crop protection in the Sahel in particular and in West Africa as a whole. Through its diverse training activities and information component, it contributes to strengthening the national plant protection services in CILSS member states. Its training programme is designed to deliver basic knowledge in plant protection to trainers to enable them to operate efficiently as field agents. The programme takes into account alternative pest control strategies, while at the same time promoting the judicious use of pesticides.

REFERENCES

ANON. (1992) *Sahel PV Info*. Bulletin d'information en Protection des Végétaux de l'UCTR/PV. No. 40.

SAGNIA, S.B. (1991) Training in entomology for middle-level technicians at the CILSS Regional crop protection training centre in Niger. *Insect Science and its Application*, **12** (1/2/3): 305-309.

Integrated control of *Striga* in Africa: the role of PASCON

S.T.O. LAGOKE, H.J. HOEVERS and S.S. M'BOOB

Pan-Africa Striga Control Network, Department of Agronomy, Institute for Agricultural Research, Ahmadu Bello University, Samaru, Nigeria

Striga species and related parasitic weeds of the family Scrophulariaceae are the most important biotic constraints to food crop production in sub-Saharan Africa (FAO, 1989). These weeds affect the livelihood of some 300 million people in Africa, causing severe damage to crops in 21 countries and moderate damage in 22 countries (Emechebe, 1991). While it is was previously estimated that about two-thirds of the 73 million ha devoted to cereal production are situated in ecological zones where these weeds are endemic, recent surveys have shown a wider distribution to areas not so far affected in the savanna and to other areas of high rainfall (Lagoke *et al.*, 1991) (Figure 1).

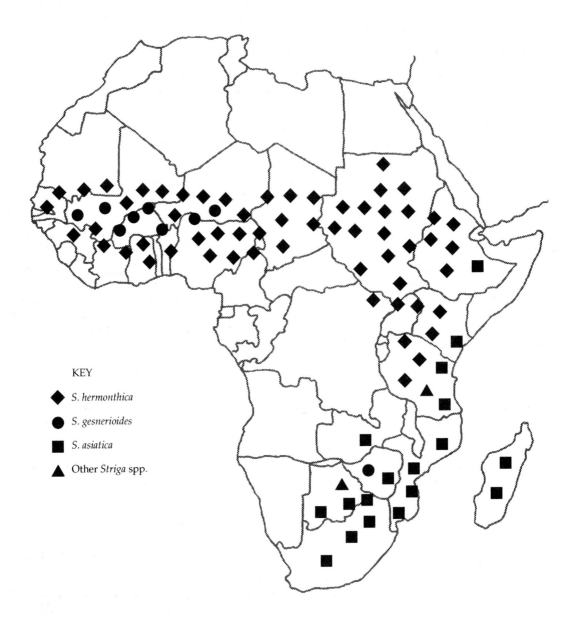

Figure 1.   Distribution of agriculturally important *Striga* species in different regions of Africa.

Some of the factors responsible for the incidence and intensity of parasitic weed infestations are:

(a) successive cultivation of susceptible crops types;

(b) the wide host range among cultivated and non-cultivated plants, such as wild grasses and legumes;

(c) the ease of distribution of the light minute seeds by wind, water and human activities, including the use of farm machinery, animal waste and contaminated crop seeds;

(d) degrading soil conditions favouring the parasites;

(e) wide genetic variability of the parasitic weeds;

(f) inadequate control of emerged weed plants, resulting in increased production of viable seeds.

Ironically, various efforts made by governments to increase food crop production and security have increased the incidence, intensity and spread of these weeds, which will easily spread to uninfected areas if adequate control measures are not taken.

Table 1. Important parasitic weeds, their hosts and distribution in Africa

Species	Host crop	Countries affected
S. hermonthica	sorghum, millet, maize, upland rice, fonio, sugar-cane, teff	Benin, Burkina Faso, Cameroon, Chad, Côte d'Ivoire, Egypt, Ethiopia, The Gambia, Ghana, Guinea, Kenya, Mali, Niger, Nigeria, Senegal, Sudan, Tanzania, Togo, Uganda
S. aspera	maize, millet, sorghum, upland rice, lowland rice, sugar-cane, fonio	Benin, Burkina Faso, Cameroon, The Gambia, Mali, Niger, Nigeria, Togo
S. asiatica	maize, millet, sorghum, upland rice, teff	Benin, Burkina Faso, Botswana, Cameroon, Côte d'Ivoire, Egypt, Ethiopia, Ghana, Kenya, Nigeria, Senegal, Sudan, Tanzania, Togo, Zimbabwe
S. gesnerioides	cowpea, tobacco, tomato, sweet potatoes	Benin, Burkina Faso, Cameroon, Chad, Côte d'Ivoire, The Gambia, Ghana, Mali, Niger, Nigeria, Senegal, Togo, Zimbabwe
S. forbesii	maize, sorghum, sugar-cane, lowland rice	Ethiopia, Nigeria, Togo, Zimbabwe
S. passargei	maize, sorghum	Mali, Nigeria
S. densiflora	sorghum	Sudan
S. klingii	sorghum, millet	Burkina Faso
S. latericea	sugar-cane	Ethiopia
S. brachicalix	sorghum, millet	Mali
Buchnera hispida	sorghum, millet	Burkina Faso, Cameroon, Mali, Nigeria
Alectra vogelii	groundnut, cowpea, soyabean bambarra nut, Dolichos velvet bean, broad bean, sesame	Burkina Faso, Cameroon, Côte d'Ivoire, Kenya, Mali, Nigeria, Senegal, Tanzania, Zimbabwe
Ramphicarpa spp.	sorghum, deepwater rice	Mali, Uganda

CROP LOSSES

Parasitic weeds derive all their requirements from the roots of the host crop. Hosts include the main staple food crops, the cereals (maize, sorghum, millet, rice, fonio, teff, hungry rice, and, more recently, wheat), grain legumes and oilseeds (cowpea, soyabean, groundnut, bambarra nut, broad bean, French bean, dolichos, green gram, black gram and sesame), and horticultural crops (tomato and potato), as well as industrial crops (sugar-cane and tobacco).

Estimated yield losses of 10–95% may occur depending on varietal reaction, ecology and agronomic practices. In 1986, crop losses in sub-Saharan Africa due to parasitic weeds were estimated at an average of 40%, equivalent to US$ 7 billion of cereals. In some areas, parasitic weeds cause complete crop failure and farmers are forced to abandon heavily infested fields in search of areas free of parasitic weeds. The resource-poor farmers, who constitute 80% of the farming population, are the most threatened by this most important biotic constraint. The greatest damage occurs in the savanna and Sahelian zones, which are the major grain producers in the region. In addition, these areas are also severely affected by other constraints, both physical (drought, low and erratic rainfall, degrading soil conditions) and biotic (other pests and diseases).

Twenty-four *Striga* species have been identified in Africa. Table 1 shows those members of the family Scrophulariaceae which constitute an immediate threat to crop production. The most economically important species in Africa are *Striga hermonthica, S. asiatica, S. aspera, S. forbesii, S. gesnerioides* and *Alectra vogelii*. Between 4 and 17 parasitic weed species are found in each country in sub-Saharan Africa (Lagoke, 1989).

CONTROL METHODS AND LIMITATIONS

Experimental work on control methods for parasitic weeds began in Africa in 1905, when the problem was first recognized. Although previous research efforts yielded some results, they did not produce an effective and appropriate solution at the farmer level. This failure may be related to the complexity of the problem, the nature and approach of past research efforts, and the available resources.

The weeds have highly specialized relationships with their hosts and exhibit wide genetic variability, resulting in different physiological strains, subspecies, ecotypes and morphotypes. They also have a wide range of prolificacy, longevity and dormancy of seeds, and some species are able to hybridize with others. This has resulted in a wide adaptation to environments and hosts, as well as different levels of virulence, making these weeds variable and difficult to control. The problem is so complex that some strategic and applied research is still required before effective control measures can be integrated. Advanced laboratories in both developed and developing countries are involved in a large programme of basic research on parasitic weeds.

Previous research efforts were confined to the development of single components for weed control, often based on the researcher's perception and discipline, rather than an integrated approach, which rarely considered the farmer's situation, the complexity of the problem and the agro-ecosystem. Many control measures developed were so specific that any slight modification in the agro-ecosystem caused a breakdown in performance. They did not fit in with the farming practices and the expected socio-economic benefits did not arise. Consequently, the farmers did not adopt these measures, whose limitations are listed in Table 2.

Table 2. Parasitic weed control measures and their limitations

Control method	Remarks/limitations
Uprooting/hand pulling	Labour and time intensive
Hoe weeding	Labour and time intensive
Land cultivation techniques	Some control of *Striga*
Crop rotation	Farmer may be obliged to grow a non-preferred crop
Intercropping	Results are conflicting and location specific
Manipulation of sowing date	Late planting of early maize and millet receives research attention
Fallow/abandoning infested field	Last, fatalistic resort
Use of organic fertilizer	Animal manure appears to be effective
Use of inorganic N-fertilizer	Contradictory research results, varying with crop varieties, location and soil conditions
Host plant	Degree of tolerance/resistance depends on the level of weed infestation
Chemical control	Still receiving research attention
Biological control	Some biological control mechanisms have potential, but have not yet been fully investigated

No one country or research organization has the resources to develop and test control packages that can be applied across the continent for the production of a wide range of crops. The National Agricultural Research and Extension Systems (NARES) lack the funds and do not have enough sufficiently trained manpower to test control technologies adequately and to develop integrated control packages. The resources of the extension and crop protection services to visit, monitor and train are too limited. This seriously hampers the transfer and implementation of available technologies to the farmers. The majority of farmers are not able to adopt or correctly use available technologies, either because the funds, manpower or supplementary inputs are lacking, or because the farm household has set different priorities. Policy-makers, therefore, need to promote the availability of the required inputs and develop an appropriate marketing system for food crops.

PASCON

Various meetings were held to draw the attention of governments, scientists, organizations and the international community to parasitic weeds as the most limiting biotic factor to food crop production in the region. The joint FAO/Organization of African Unity (OAU) All Africa Government Consultation on *Striga*, held in Maroua, Cameroon, in October 1986, was the first effective attempt to provide an inventory of knowledge, activities and findings of individual institutions and nations, and to assess their effectiveness to provide a solution to the problem. The conclusion of several international workshops on *Striga* since November 1983 was that the various efforts were being made in isolation, with little exchange of information, and without the benefit of a co-ordinating framework.

Given the pan-African nature and seriousness of the problem the need for a concerted, co-ordinated regional programme was evident. It was the recognition of this critical deficiency and the need to avoid duplication of effort and wastage of resources that led to the creation of the Pan-Africa Striga Control Network (PASCON), in Banjul, The Gambia, in December 1988. The objective, structure, governing entities and research agendas were considered in detail at the first PASCON-FAO meeting in Ibadan, March 1990.

Objective and organization

The purpose of PASCON is to provide a co-ordinated system for mutual collaboration and co-operation among various national programmes, and between them and international/regional agricultural institutions, in parasitic weed research, especially control.

The specific objectives of the network are:

(a) to promote collaborative research on control methods of parasitic weeds and their application;

(b) to strengthen the capabilities of NARES in the areas of parasitic weed research and control;

(c) to facilitate exchange and dissemination of information on parasitic weed research and control among participating institutions within and outside the region.

Membership of PASCON is open to:

(a) NARES involved in all levels of parasitic weed research, control and extension in sub-Saharan African countries;

(b) International Agricultural Research Centres (IARCs) and advanced laboratories involved in strategic research on parasitic weeds and their control;

(c) Regional Research Co-ordination Agencies (RRCAs) associated with parasitic weed research or with the network.

The numbers of NARES participating in PASCON has increased from 4 to 18 comprising Benin, Burkina Faso, Burundi, Cameroon, Ethiopia, The Gambia, Ghana, Kenya, Mali, Niger, Nigeria, Rwanda, Senegal, Sudan, Tanzania, Uganda, Zaire and Zimbabwe.

All participating institutions and agencies meet every two years at the General Workshop. The General Workshop reviews and assesses on-going activities, identifies research needs and priorities, and promotes the exchange of technical information, research methodologies and experiences. The Steering Committee is the driving force of the network. It comprises research scientists of participating NARES, IARCs and advanced laboratories, elected at the General Workshop.

General Workshops

Since the establishment of the network two workshops have been held, the first at IITA, Ibadan, March 1990, and the second in Nairobi, June 1991.

Delegates to the first General Workshop reviewed the activities of the network, exchanged experiences, and planned future activities. They also discussed research methodologies, constraints and results, and expressed the need to standardize research protocols and survey methodologies for effective exchange on the *Striga* problem and develop control packages for different situations. Delegates recognized that on-farm research was essential and appropriate to national plant protection research institutions. The need for basic research was also emphasized and found appropriate to international centres, and long- and short-term research priorities were identified (Table 3). Emphasis was placed on the need for on-farm testing of packages based on existing research information on available components (short-term), and on those developed by national programmes (long-term).

Table 3. Long-term and short-term research priorities for parasitic weed control

Long-term
(a) Reduction of weed seedbank in soil, seed ecology, host range and host specificity
(b) Effects of fertilizer, soil structure and soil microbiology on parasitic weeds in crops
(c) Effects on parasitic weeds of trap crops in rotation with or without mixed cropping; breeding for resistance and tolerance
(d) Development of crop loss assessment techniques
(e) Socio-economic implications of the parasitic weed problem
(f) Development of synthetic germination stimulants
(g) Mechanisms of resistance to, and tolerance of, parasitic weeds in crops and characterization of stimulants produced by host crops
(h) Biocontrol measures

Short-term
(a) Identification of herbicides that would be useful for pre-emergence control of parasitic weeds
(b) Screening available post-emergence herbicides and evaluating them for effectiveness
(c) Methodologies of screening for selection of resistant and tolerant varieties
(d) Extent of contaminations of seed lots with parasitic weed seed
(e) Development of appropriate scoring scale for assessment of weed infestation

The second General Workshop, funded by FAO, was attended by 89 participants representing 16 national programmes, the IARCs, Inter-African Regional Organizations, advanced laboratories, technical agencies (FAO), as well as participants at the Fifth International Symposium on Parasitic Weeds. The two workshops have promoted effective interaction and exchange of information, experience and ideas among scientists of NARES, IARCs, advanced laboratories and RRCAs. A third General Workshop is proposed for March 1993.

Activities

The major activities of PASCON are working towards the implementation of effective IPM programmes (NRI, 1991). These include the promotion and strengthening of:

(a) strategic and applied research and surveys to identify the levels of infestation, crops affected, feasibility of control activities and perception of farmers to relate the parasitic weed problem to the farming system, and to identify existing control methods and their effectiveness, as well as the generation and evaluation of control components;

(b) the development of integrated packages for on-station and on-farm testing;

(c) on-farm/pre-extension evaluation and adaptation of control packages;

(d) extension of proven control packages.

Other activities include information dissemination and exchange through the newsletter, proceedings, reports of meetings, inventories and workshops, and the strengthening of NARES through the training of scientists and provision of, or negotiation for, resources where necessary. In spite of the limited funds, currently borne by FAO, PASCON has successfully undertaken some of these activities.

Research

The FAO parasitic weed control projects in Cameroon and The Gambia have recently ended. They generated useful approaches to combat the *Striga* problem. Seven national programmes with adequate infrastructures and experienced manpower have been identified for the on-farm evaluation of integrated packages for parasitic weed control. The centres have the technology but not the funds to carry out the necessary field work. It is hoped that PASCON will help find financial assistance from other sources.

Visits have been made to various national programmes to provide technical support and to assist in preparing proposals for research activities. Exchange of information and technical resources, e.g. germplasm and herbicides, were enhanced.

Information exchange

The *Striga Newsletter*, the network's major instrument of information exchange, is widely distributed to all participating scientists and extension specialists. Two issues have been published, and a third issue is in preparation. The newsletter has attracted requests from both within and outside the region. The reports of the first and second General Workshops have been published and distributed. Directories of parasitic weed workers listing their research activities are being prepared for circulation.

Through a training workshop organized and funded by IITA a parasitic weed research manual for maize was produced. A more comprehensive manual for all crops is being prepared by the network, together with an inventory of control technologies.

Training

Through PASCON, NARES staff have been placed in institutions for further studies, and given short attachments at some IARCs. A training workshop, organized and funded by IITA in October 1990, was attended by scientists from IITA as well as five scientists from NARES in West and Central Africa. A similar workshop is planned for October 1992. This workshop will be organized jointly by PASCON and FAO, and financed through FAO's Technical Co-operation Programme. It will be attended by scientists from 17 countries participating in PASCON.

Co-ordination

PASCON has emphasized the co-ordination and integration of parasitic weed activities in each country, and promoted the establishment of nationally co-ordinated groups to avoid duplication of effort and waste of resources. Such groups have been formed in Cameroon, Ghana, Kenya, Mali, Nigeria, Tanzania and Uganda.

The network interacts with other RRCAs, such as RENACO (cowpea), EARSAM (sorghum and millet), and WCASRN (sorghum). Possible areas of collaboration with PASCON and the national programmes, and in particular the SAFGRAD-PASCON inter-network relationships, are discussed in meetings with regional organizations, such as CILSS and SAFGRAD. Regular meetings were held with IARCs (IITA, International Crops Research Institute for the Semi-Arid Tropics (ICRISAT), International Maize and Wheat Improvement Centre (CIMMYT) and International Centre for Insect Physiology and Ecology (ICIPE)) to ensure their participation in PASCON activities.

Integrated control packages

PASCON was set up following the review of FAO projects and national programmes that evaluated some integrated packages for parasitic weed control. Farmers have always combined their local tolerant or resistant varieties with two or more cultural practices, e.g. tillage, hand pulling, uprooting, hoeing, crop rotation, fallow, and the use of fertilizer to ensure the production of an acceptable yield of food grain crops under weed infestation.

The main objectives of integrated weed control packages are to reduce (a) crop yield loss, and (b) seed production to prevent build-up of the weed seed population in the soil.

The development of integrated packages involves the combination of proven and effective components to ensure that they are technically reliable, adaptable, socio-economically acceptable and sustainable. It is essential that farmers are consulted during the development of the package and that multi-location testing is done to ensure adaptability to various agro-ecosystems.

Preventive measures, such as the use of clean farm equipment, seeds and manure that are not contaminated with weed seeds, and the destruction of emerged weeds before flowering, could be simulated and implemented through extension services and legislation.

Varieties which combine resistance or tolerance to parasitic weeds with acceptable grain quality and agronomic performance are the cheapest and best component of integrated control packages.

Manual removal of weeds can be used with other methods for light infestations but is more difficult and requires multiple operations if the infestation is high.

Organic and inorganic fertilizer enhance the tolerance of both cereals and grain legumes, but may not necessarily reduce infestation. Nitrogen at 80–100 kg/ha for sorghum and 90–150 kg/ha for maize has been reported to improve crop performance. Nitrogen also reduces *Alectra* and *Striga* in cowpea but depresses yields, while phosphorus increases yield and *Alectra* infestations.

Pre-emergence herbicides reduce weed competition and early damage, while post-emergence herbicides reduce further crop loss and prevent seed production by weed plants by destroying them.

Trap crop and non-susceptible hosts like soyabean, cotton, cowpea and bambarra nut can be grown in rotation or in association with cereals to reduce weed infestations. Millets can also be used in certain ecological zones where the weeds parasitize other cereals. It is, however, essential that the selected crops fit into the existing farming system.

These specific components can be modified as other more promising components become available.

Achievements

PASCON has increased the awareness of the parasitic weed problem at various NARES, IARCs and RRCAs so that parasitic weed activities have become priority projects. These organizations have also set up working groups to tackle the problem. Collaboration has improved and PASCON is recognized as the focus for all parasitic weed activities in the region. Three countries have officially released recommendations for parasitic weed control in maize (Nigeria) and sorghum (Sudan, Niger and Nigeria). Many other countries have produced packages for various crops, including cowpea, that are being tested before they are released officially.

Constraints

The major constraint preventing a wider network is the lack of funds. Most of the participating countries have local, qualified scientists to carry our parasitic weed research, but lack the necessary financial resources to carry out field experiments. These national programmes need financial assistance of about US$ 5000 annually to conduct on-farm trials. The parasitic weed problem is endemic and could become more serious if effective control packages are not developed and tested for adoption by farmers.

Given the present economic difficulties facing many African governments, PASCON will have problems in generating income on its own, or in becoming fully funded by the member countries, at least in the foreseeable future. FAO is actively seeking donor support for the activities of the Pan-African *Striga* Control Network.

In conclusion, PASCON wishes to thank FAO for the financial, technical and logistic support it has enjoyed since its inception, and the sponsorship enjoyed by PASCON membership for various activities from other agencies such as the International Development Research Centre (IDRC), USAID, EC, Canadian International Development Agency (CIDA), SAFGRAD, CILLS, Southern African Development Co-ordination Conference (SADCC) and EARSAM.

The technical support by IARCs (IITA, ICRISAT, CIMMYT and Centre de Coopération Internationale en Recherche Agronomique pour le Développement (CIRAD)) and many NARES, and the co-operation of various IARC and NARES scientists in ensuring effective co-ordination is highly commendable.

REFERENCES

EMECHEBE, A.M. (1991) *Pan-African* Striga *Control Network (PASCON): Feasibility and Relevance.* FAO-Regional Office of Africa.

FAO (1989) *Striga* – improved management in Africa. Proceedings of the FAO-OAU All Government Consultation on *Striga* Control, Maroua, Cameroon, October 1986. *FAO Plant Production and Protection Paper*, 96. Rome: Food and Agriculture Organization.

FAO (1991) *Striga* in Africa. *Second PASCON Workshop, June 1991, Nairobi.*

LAGOKE, S.T.O. (1989) Striga *Situation and Control in Africa.* Programme of action for combating the *Striga* problem in Africa. Report of consultancy to IAPSC/STRC/OAU, Yaoundé, July, 1989.

LAGOKE, S.T.O., M'BOOB, S.S., HOEVERS, H. and EMECHEBE, A.M. (1991) Combating *Striga* problem in Africa: Pan-African *Striga* Control Network Approach. *Thirteenth Conference of the Weed Science Society for East Africa, October 1991, Nairobi.*

NRI (1992) *A Synopsis of Integrated Pest Management in Developing Countries in the Tropics.* Chatham, UK: Natural Resources Institute.

Successful implementation of IPM for tree crops in West Africa

A. YOUDEOWEI

West Africa Rice Development Association (WARDA), 01 B.P. 2551 Bouaké, Côte d'Ivoire

INTRODUCTION

This paper presents some successful experiences of IPM implementation on tree crops in West Africa. A discussion of the issues concerned with this success must consider two important factors.

Management of the major pests of tree crops has been closely linked with maintaining high yields from economically valuable crops such as cocoa, coffee, oil palm, rubber and mango. Since this is a substantial source of revenue for both governments and farmers this is a very sensitive area.

So far there has only been limited 'success' with the implementation of IPM on tree crops; it is more realistic to recognize that any claims for IPM success have been made in the context of positive initiatives towards IPM implementation.

This paper will discuss the main technical, socio-economic, institutional and political issues which have been important in the implementation of IPM on tree crops in West Africa. Cocoa and mango will be used to illustrate the main issues under consideration.

COCOA

The best environment for cocoa farming in West Africa is well understood in relation to the ecological changes which influence the dynamics of the major pest and disease organisms. Cocoa farmers are knowledgeable about production systems and are receptive to new crop protection and other technologies. Cocoa is an important economic and political crop, farm gate prices and production inputs are common issues in political campaigns. Because of this, research and training activities, especially those concerning the management of major pests and diseases, have received considerable political and financial support from governments in West Africa, notably in Nigeria, Ghana and Cameroon which are the major producers in the sub-region.

To evaluate the success of IPM in cocoa, the following issues must be considered.

Technical issues

Extensive research over the past 50 years has helped to identify and develop components of an IPM protocol for the management of the major cocoa pests and diseases. These include:

- agronomic practices, such as crop canopy management and farm sanitation;

- pest identification and regular surveys of the incidence and spread of pests and diseases in relation to changes in the ecological conditions on cocoa farms;

- spot chemical treatment at early stages of pest and disease infestation—this drastically minimizes the routine and blanket application of pesticides which has detrimental effects on the cocoa environment, especially to beneficial arthropods in the ecosystem.

These components have been used with limited success in the management of cocoa pests and diseases. Greater success could be achieved if the constraints which influence the more effective implementation of IPM were removed. A critical multidisciplinary study and evaluation of farmer behaviour in the choice of pest management tactics at different times in the crop cycle is needed to understand the factors influencing the adoption of IPM technologies.

Socio-economic issues

The declining economic situation and the removal of government subsidies on agricultural inputs, together with the devaluation of local currencies in Nigeria and Ghana, have raised the costs of many pesticides well beyond the reach of farmers. Cocoa farmers have been forced to adopt cheaper pest management methods, which are fortunately environmentally safer to deal with pest problems. This is an ideal situation for the implementation of IPM technology.

26

Institutional issues

The establishment of cocoa research institutes in West Africa has provided facilities for research on pest management. This has contributed substantially to the promotion and implementation of IPM in the cocoa agro-ecosystem.

The major constraint to success in IPM implementation is the lack of co-ordinated multidisciplinary team research on IPM strategies for the major cocoa pests and diseases. Research should focus on integrating pest management tactics at farmer level. Involvement of farmers at the design and implementation stage is needed for more relevant and effective research.

Other important factors are education and training of farmers and extension personnel on IPM technologies using audio-visual aids and other techniques to disseminate information and promote IPM technology.

Political issues

The economic importance of cocoa as a major export earner of foreign currency has encouraged governments in West Africa to provide political and financial support for cocoa research from which IPM has benefited directly. So far there has been no direct political support for IPM on cocoa. When it becomes available political support should be directed in two main areas:

(a) the establishment of national policies on IPM which would lead to the allocation of adequate resources for the development and promotion of IPM;

(b) to set up farmers' groups to implement IPM on cocoa and other tree crops.

MANGO

The mango mealybug, *Rastrococcus invadens*, was recognized as a major pest of mango in Nigeria in 1987. It was first reported in Lagos at the State House premises. Since then it has spread all over the country and now infests and destroys several other tree and horticultural crops. It is also a pest in neighbouring West African countries.

The management of *R. invadens* in Nigeria by the adoption of IPM has been successful for the following reasons.

(a) The Nigerian government declared the mango mealybug a national pest and thus gave political recognition to it.

(b) A research institution, the National Horticultural Research Institute (NIHORT), was given the responsibility for developing IPM technology to deal with this pest problem. The government provided adequate financial resources to NIHORT for this task.

(c) NIHORT established a survey and research programme on the mango mealybug, in collaboration with scientists from the Universities of Ibadan and Ife, the National Plant Quarantine Service and the IITA Biological Control Centre for Africa. Technical sub-committees were set up to deal with specific components of IPM, such as pest surveillance, cultural control, biocontrol, chemical control and legislative control.

(d) Through these efforts, a co-ordinated IPM strategy for the mango mealybug was developed and implemented.

The main point illustrated by these two examples is that the successful implementation of IPM on major crop pests is possible through government intervention. In these cases the governments provided political and financial support for research, promotion and implementation of IPM strategies on major pests.

Discussion

Much of the discussion following this session concerned issues of training in plant protection raised by S.B. Sagnia. Mr Sagnia was asked to elaborate on the background of the training institute and its students. He explained that in order to be admitted to the institute students must possess a baccalaureate and at least two years field experience. The first year of the course included review and upgrading to ensure that all students possessed a sufficient understanding of the basic sciences, including plant pathology and entomology. At present placing graduates is not a problem, most took up posts as extension agents, research assistants and teachers. Since the course has received official recognition, graduates are also able to pursue further studies elsewhere in Francophone Africa as well as abroad.

A. Latigo asked if the training programme included any evaluation mechanisms and if training was extended to the farmer level. Mr Sagnia responded that a follow-up of former students and their employers was carried out every two years and that continuous contact with former students was maintained through a quarterly newsletter. Based on feed-back from former students, the institute has introduced new course components, such as an extension component, and provided opportunities for upgrading.

P. Matteson remarked that studies from the Philippines, Indonesia and Sri Lanka have shown that the Training and Visit model of extension offered by the institute is, in fact, poorly suited to IPM extension. It has been found that new innovative forms of extension based on informal adult education techniques and group approaches have been much more successful. Experience has also shown that evaluations must be extended to farmers themselves if they are to be accurate and effective.

S. M'Boob questioned the sustainability of training institutes such as that described by Mr Sagnia. He stressed that national governments must assume responsibility for the funding of training programmes and asked if any thought had been given to the future of the institute at such a time when external donor funds were no longer available. Mr Sagnia responded that the training institute had a goal of providing at least 5% of its own funding and planned further attempts to gain funds from home governments.

A. Youdeowei cited the example of a master's programme in plant protection recently introduced in Morocco, as well as initiatives in the development of training for trainers programmes and training aids. He asked Mr Sagnia if his training institute had given any thought to these areas. Mr Sagnia responded that the development of training aids had been initiated but that progress was limited by lack of funds. It was hoped that training for trainers could be extended in the future, especially for the training of part-time teachers not necessarily engaged by an agricultural college or training centre.

Commenting on S. Lagoke's presentation of PASCON, Mr M'Boob observed that considerable progress had been made since the inception of the network and that few funds should now be required to ensure its continued existence. When asked why FAO was discontinuing its financial support of PASCON, Mr M'Boob replied that FAO planned to continue its technical support to the network, but not being a funding agency did not possess the necessary funds to maintain the network financially. Representatives from IITA expressed surprise and concern at the FAO's claimed lack of access to funds, and expressed the hope that given its position and influence within the UN system, FAO could use its leverage to mobilize continued donor support.

With respect to A. Youdeowei's presentation, a representative from Togo commented that financial gain appeared to be the most important criterion for consideration when working with farmers. Experience with cocoa farmers in Togo has shown that farmers would do anything to protect their cocoa, but that they had little interest in other crops. Farmers only seemed interested in crops of high economic value and attempts to introduce programmes related to less valuable crops are not successful.

Development of the Haute Vallee (DHV) project

A. DREEVES

DAI/HVV Project, United States Agency for International Development (USAID), Bamako, Mali

The Development of the Haute Vallée (DHV) project in Mali is designed to assist farmers and extension workers with technical support, training and education in appropriate and economical pest management using integrated approaches. In 1992, the USAID-supported Operation Haute Vallée (OHV) project focussed on farmer participation in learning technology at village level with hands-on teaching and implementation, and training.

Education is very important for the successful implementation of alternative low-input sustainable agriculture and is essential to encourage farmer participation. This programme aims to identify the constraints to implementing IPM.

The programme will aim to: change attitudes; bring out a new awareness; sensitize all levels of the public; seek a commitment from the government; and link gaps.

The programme is divided into four components.

1. Training extension agents and farmers through visual demonstrations.

Three modules (implemented with the aid of a three-sided display board)

(a) Steps to successful pest management:
- **principles** for effective IPM
- **strategies** – Options – What is IPM?
  – How can it work?

(b) Use common sense, look closely, know and understand your field. Be a responsible pesticide user:
- shortcomings and dangers of pesticides
- choosing chemicals (justification in particular cases)
- choice, types, user safety, storage, disposal, first aid.

(c) On the road to agricultural self-sufficiency:
- selection of 10 farmers by farmers
- identify economical and low-input options/practices
- criteria for choosing options.

2. Farmer involvement on farm field day.

Key components:

(a) 'Learning by doing' approach.

(b) Collection and identification of problems.

(c) Investigation of knowledge, attitudes and practices.

(d) identification of 3–4 approaches to install and implement as tactics, i.e.
- soil preparation techniques
- planting schemes: intercropping with sorghum and cowpeas
- sampling/monitoring system
- problem solving.

3. Promotion of IPM education through media.

Incentives
- certificate of farmer appreciation
- recognition of farmer participation – a posting in field.

Radio release/newspaper on IPM.

Promotional brochure on IPM.

Video with simplified messages.

4. Evaluation/assessment (Pre- and Post-)

Did the programmes approach through training with posters and the media reach the extension workers and farmers in a way that we wanted? Questions/survey to evaluate how things were perceived.

The CILSS integrated pest management project (1980–87)

D.D. BÂ

CILSS, B.P. 1530 Bamako, Mali

The CILSS Regional Integrated Pest Management Project, financed by USAID was designed to run over three, 5-year stages. However, only the first stage has been implemented. During this stage the project received technical support from FAO. The first two years were mainly devoted to establishing infrastructures and acquiring equipment. At the end of the first stage, USAID took the decision to provide support to Sahelian states on a bilateral basis.

AIMS AND OBJECTIVES

The long-term objective of the IPM project was to help increase the agricultural output of foodstuffs by reducing losses caused by insects, diseases and weeds. The aim was to use effective, rational and economic control techniques which were appropriate to the environment and which would minimize the use of chemicals. This was to be achieved by paying particular attention to the forecasting of attacks and giving warnings for the control of pests.

These aims were to be achieved through:

(a) the establishment or strengthening of research centres to investigate the bio-ecological complex of the main enemies of the major food crops in order to develop integrated control methods;

(b) the training of personnel at all levels in research, surveying, reporting and IPM techniques;

(c) the establishment of a monitoring system for major crop pests;

(d) the identification of economically important pests and determination of the losses which they cause to crops;

(e) analysis and evaluation of traditional methods of crop protection with a view to possible improvement;

(f) the provision of demonstration plots for the application of research results to assess the receptiveness of farmers to new or improved technologies, and demonstration of potential benefits.

STRUCTURE AND ORGANIZATION

The regional project director was assisted by an FAO technical counsellor. The function of the regional directorate was, essentially, to provide technical co-ordination for the programmes. The directorate also included a bioclimatology unit and a socio-economic unit.

A national division was set up in each CILSS country. This was led by a director, appointed by the national institution with responsibility for the project, assisted by an FAO principal expert. Each participating body was responsible for implementing the project at national level.

In each country, the national working party, comprising the national director, the FAO principal expert and the USAID project liaison officer, was responsible for ensuring that the project was carried out satisfactorily. At the regional level, a project working party had similar responsibilities. A three-part technical committee, consisting of the executive secretary of CILSS, and representatives of USAID, Washington and FAO, Rome, met once a year to examine and adopt annual programmes and their corresponding budgets.

A national co-ordinating committee, consisting of representatives of agricultural research, the national plant protection service and the department of agriculture, was responsible in each country for co-ordinating the project with national programmes and with bilateral or multilateral aid.

The project was responsible for building and equipping 11 research laboratories (Cape Verde 2, Burkina Faso 3, The Gambia 1, Mali 1, Mauritania 2, Senegal 2), fitting out an office-laboratory complex in Chad, building offices (Burkina Faso 1 and Niger 2) and housing research personnel in Cape Verde and The Gambia. In addition, the project fitted out, or provided additional equipment, for six national laboratories (Cape Verde 1, The Gambia 1, Senegal 2, Niger 2). The project also constructed eight insectariums (Burkina Faso 2, Cape Verde 2, The Gambia 1, Mali 1, Niger 1, Senegal 1) and five glasshouses (Burkina Faso 2, The Gambia 1, Mali 1, Senegal 1) for particular research purposes.

This well-equipped infrastructure should facilitate the development of IPM at national level.

THE ESTABLISHMENT OF A MONITORING SYSTEM

Observation posts manned by project technicians were built or fitted out in crop areas, in relation to their ecology. These units were provided with agro-metereological equipment, field optical equipment and biological handling equipment.

As many as 55 such units were provided throughout the Sahelian growing area: 11 in Burkina Faso, 6 in The Gambia, 7 in Mali, 12 in Mauritania, 11 in Niger, 4 each in Senegal and Chad. In Cape Verde the project used local plant protection centres.

The observation units, located within the farming environment in the main agricultural ecological areas, provided a light and functional framework with low running costs for a pest-monitoring system for which the development of an agricultural warning system would have been unsuitable. The availability of well-trained and experienced personnel will make it possible to consider extending this network to those major agricultural production areas which are not easily accessible.

PERSONNEL TRAINING

The project granted 31 bursaries for university training at second and third levels. Of these, 27 individuals gained diplomas comprising 4 for Burkina Faso, 2 for The Gambia, 7 for Mali, 7 for Mauritania, 2 for Niger, 3 for Senegal, 1 for Chad and 1 socio-economic diploma for the regional component. At the completion of the project, the trained socio-economist from Burkina Faso went back to his own country.

National and international research workers provided basic training in integrated management for 125 national technicians who were attached to the project in laboratories and observation posts. This training was supplemented by an annual two-week refresher course in each country.

By the end of the first stage of the project there was a pool of skilled individuals at national level, both technicians and research workers, to carry out IPM work.

RESEARCH PROGRAMME

The project concentrated on millet, sorghum, maize, rice, cowpea and groundnuts; other minor food crops were excluded. An exception was made in the case of Cape Verde, where maize, and to some extent cowpea, are the only crops grown. Here, in addition to maize and cowpea, the project's included other legumes and cassava.

During the first two years of the project eight pests of economic importance throughout the Sahel were identified:

- the millet head-miner (*Heliocheilus albipunctella*)
- the sorghum midge (*Contarinia sorghicola*)
- millet downy mildew (*Sclerospora graminicola*)
- millet smut (*Tolyposporium penicillariae*)
- sorghum smuts
- *Striga hermonthica* affecting millet and sorghum
- *Striga gesnerioides* affecting cowpea
- rice blast (*Pyricularia oryzae*).

Additional observations showed that the sorghum midge was not of regional economic importance and that another group of insects, the meloid beetles (*Psalydolytta* spp. and *Mylabris* spp.) were restricting millet production in several regions of Mali, Mauritania and The Gambia.

31

Methods of study and standard experimental protocols were developed for those pests which were affecting several countries, so that the results could be used more effectively. In addition, pests of food crops which were of geographically limited economic importance were identified and were investigated at national level.

The results of research into these various pests carried out between 1983 and 1986 are available from CILSS (Sahel Institute). Once the project ended, investigative work continued at national level, and the results obtained between 1983 and 1991 are also available from the Sahel Institute.

BIOCLIMATOLOGY

The project included a bioclimatological component whose functions were:

(a) to contribute to the choice of sites for the observation posts forming part of the monitoring system;

(b) to equip the observation posts with standardized meteorological instruments, developing a common procedure for observations, and codifying the observation system;

(c) to collect and check bioclimatological data from the observation posts;

(d) to analyse bioclimatological data on a regional scale for use in the provision of warnings to agriculturalists.

A bioclimatological data collection system was set up through the observation posts equipped for the purpose. The observers were instructed in methods for the collection of this data.

A bio-ecological computer model for predicting the growth of grasshopper populations was developed and improved, and made available to the national plant protection services. Evaluation of its performance during the 1986 season revealed weak points in the model, which was improved later.

A computerized bio-mathematical model simulating the growth of populations of *Heliocheilus albipunctella* was prepared. Its correlation with observations made in the field was to have been improved as data were collected during the second stage; however, the project was stopped at the end of the first stage.

PILOT PROJECTS

As the project obtained confirmed results for millet, and promising results in the case of other crops, it was appropriate that the results for millet should be disseminated among the farmers, and that other results should be tested for their acceptability. Pilot activities were initiated within the farmer environment, representing a first link in the chain of extension of IPM methods.

The choice of villages and farmers was based on a number of criteria—essentially the aim was to provide reliable information which would spread to farmers in the surrounding area. Farmers for the pilot projects were chosen by an assembly of farmers, including women. Cultivation consisted of two plots of approximately 0.5 ha each, one being farmed with conventional practices (which might be traditional or improved), the other with practices recommended by the research, and including different methods of control.

Cultivation using the techniques in the pilot project produced yields 34–95% greater than traditional cultivation, depending on individual villages.

SOCIO-ECONOMIC FACTORS

Awareness of socio-economic factors involved:

(a) analysing and understanding traditional methods of crop protection, so that they might be used, after improvement if necessary, within the framework of IPM;

(b) evaluating the compatibility between IPM methods and the social, agricultural and economic characteristics of the farmers and the country.

A survey was made of traditional methods of pest control. These were evaluated sociologically and technically; some were considered to be acceptably effective and were tested. Some proved to be useful and were included in the recommendations for methods of control, while others were rejected.

As part of the pilot project, the following socio-economic constraints were surveyed.

(a) Economic calculation: the difficulty here lies in fixing the reference price of millet. Given that there is no organized market, actual prices fluctuate in relation to volume of production, and the time and place of the transaction.

(b) Agricultural constraints: working times for correct preparation of the soil, sowing in rows, weeding at a suitable time, correct evaluation of the quantities of inputs required (percentages for seed disinfection products, estimation of cultivated areas), and thinning out to two or three plants.

(c) Economic constraint: availability of finance for purchasing products other than seed disinfectant.

(d) Organizational constraints: supply of materials at the appropriate time and inadequate packaging.

(e) Social constraint: reluctance of farmers to change from conventional practices.

Particular mention should be made of the farmers' reluctance to undertake thinning, despite the fact that this has been recommended for several decades. After seeing the results, however, the farmers were convinced of the validity of this practice. This emphasized the essential role of extension, particularly the use of demonstration fields combined with appropriate information.

SCIENTIFIC MEETINGS

The project held a meeting for research workers each year between 1983 and 1987, at which the results of the national components were examined and working programmes for the following year were discussed. In December 1984 the project organized the first international seminar on IPM of food crop pests in the Sahel.

REGIONAL CO-OPERATION

Following USAID's decision to discontinue support, the Sahelian countries decided to continue co-operating on a regional basis. In January 1987 the Council of Ministers of the CILSS decided to set up the Unité de Coordination Technique Régionale en Protection des Vegetaux (UCTR/PV) based at the Sahel Institute.

The function of the UCTR/PV is to monitor the progress of research for integrated control, building on the co-operation which was built up between the Sahelian countries. As part of this monitoring process, the UCTR/PV has organized a series of scientific meetings, as follows.

(a) An annual meeting of research workers to examine the results of work carried out since the end of the project, as well as proposed programmes for the following agricultural season. The purpose of the meetings is to encourage research towards IPM within each country and to enable the research workers to co-operate—for example, in the exchange of harmonization of test protocols, and exchange of seeds of resistant or tolerant varieties.

(b) An annual meeting of the directors of the national plant protection services at which a summary of recent research is presented to those responsible for plant protection to enable them to direct their programmes towards IPM.

(c) A meeting in 1992 to bring together research workers and the directors of the national plant protection services. The aim was to improve co-operation between research workers and extension workers by making researchers more aware of the practical difficulties experienced by farmers.

(d) An annual reunion of the heads of the plant health bases (decentralized structures) in sensitive frontal areas has been set up to assist co-operation in the field and orientate control towards IPM.

(e) The second international seminar on IPM in the Sahel was organized in January 1990. This seminar brought the research carried out up to date and determined priority research themes for the following five years. As part of these themes, 11 projects were defined and grouped into the Sahelian IPM system. A main research centre and secondary centres are envisaged for each project.

(f) An international symposium on the losses resulting from pests was organized in March 1991.

The UCTR/PV publishes:

(a) an information bulletin on plant protection, *Sahel PV Info*, published 10 times a year;

(b) an annual report on plant protection activities in the Sahel;

(c) brochures;

(d) proceedings of meetings and the international seminar.

Discussion

The discussion focussed mainly on the disbanding of the IPM project described by D.D. Bâ. A representative from Nigeria asked whether a feasibility study had been carried out, and a representative from Ghana asked for further details on the discontinuation of the project. It was revealed that feasibility studies carried out in 1976 led to five project recommendations, one of which was the IPM project in question. Mme Bâ attributed the decision to end the project to a change in donor priorities away from regional projects towards those funded bilaterally. A country representative from Niger pointed out that the project has continued on a bilateral basis in a number of countries. In Niger, additional funds were sought from other donors, and the project continues with considerable support from the United Nations Development Programme (UNDP).

W. Knausenberger from USAID clarified numerous issues concerning the funding and disbanding of the project in question (see page 58). These days most donors preferred to fund bilateral projects since large-scale, broad-based, comprehensive projects had seldom met with much success. He stressed that smaller-scale research grants at the institutional or national level were a more likely prospect, and that when seeking funding, national institutions must aim to produce well-defined proposals which clearly outline IPM needs and goals. He added that more and more donors were faced with increasing demands and decreasing resources and that people could expect to see a higher level of 'focus and concentration', as opposed to broad-based funding, on the part of many donors. In the case of USAID, for example, he pointed out that a decision had recently been taken to channel 80% of African funds to 19 priority countries, rather than the 40 or so currently receiving funding.

NGO PERSPECTIVES

NGO support for crop protection in Africa

A.M. DIOP
Rodale International, B.P. A237 Thies, Senegal

The Rodale Institute is a non-profit-making organization based in Pennsylvania, United States. The international division co-ordinates the activities of two resource centres for regenerative agriculture, one in Africa, Centre de Regeneration des Resources Agricoles (CRAR)-Senegal, and the other in Central America, CRAR-Guatemala. It also co-ordinates co-operation with universities and other partners.

Regenerative agriculture is a system which can improve the basic resources of a farming activity and increase its productivity and profitability, whilst preserving the natural environment. The aim of Rodale International is to support communities at the grass-roots level to plan, experiment and implement appropriate regenerative techniques, to increase their resources and production capacities and assume effective control over their own development.

To achieve this, Rodale Senegal has drawn up a programme with the following components:
- communication and exchange of information
- research activities
- community activities and training.

COMMUNICATION AND EXCHANGE OF INFORMATION

The Rodale Senegal team established and runs an information collection and dissemination unit (Figure 1). This is at the disposal of various development agencies for the dissemination of their experiences, successes and the exchange of information on research and development activities. The main tool for this exchange is the newsletter on regenerative agriculture called *Entre Nous* (*Between Us*). One section of this newsletter deals with natural crop protection which is a component of pest management.

RESEARCH ACTIVITIES

The term 'IPM' is more comprehensive than the term 'integrated control methods'. Good crop management should be a part of the technologies intended to reduce and minimize parasitic activity and crop pests. Three aspects of these technologies are currently being studied with active farmer participation:
- use of compost
- mixed cropping
- natural crop protection with the help of plant extracts (e.g. neem)

Two main constraints have been identified:
- impact certification and evaluation
- marketing.

COMMUNITY ACTIVITIES AND TRAINING

The active participation of the rural communities to identify problems and draw up programmes of activities and evaluation should be encouraged. In order to fulfil this objective, the training of trainers and of the rural communities is essential. A farmers' seminar was organized jointly with the Institut Senegalais de Recherches Agronomiques (ISRA) in 1991, funded by USAID. This was successful in involving farmers and establishing a network of researchers, NGO/extension services and farmers. The centre is also currently developing close co-operation with state institutions and other organizations like the US Peace Corps.

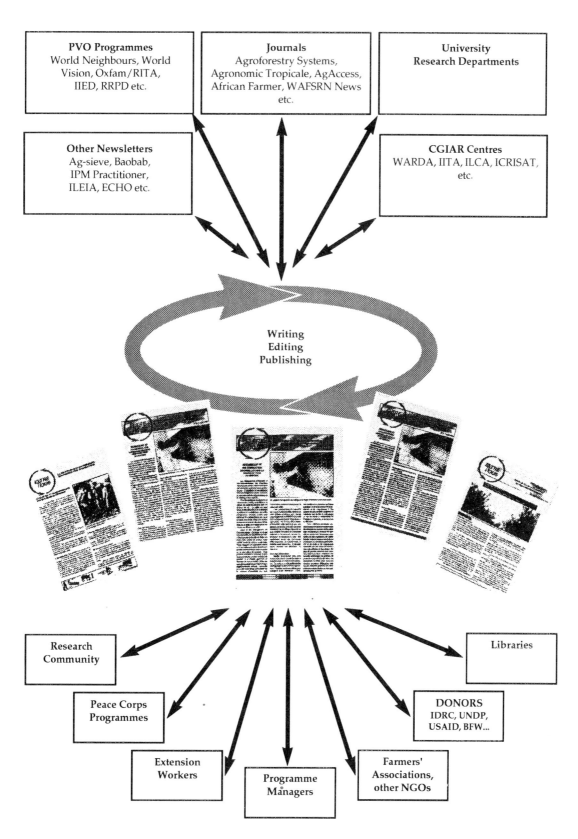

PVO Programmes
World Neighbours, World Vision, Oxfam/RITA, IIED, RRPD etc.

Journals
Agroforestry Systems, Agronomic Tropicale, AgAccess, African Farmer, WAFSRN News etc.

University
Research Departments

Other Newsletters
Ag-sieve, Baobab, IPM Practitioner, ILEIA, ECHO etc.

CGIAR Centres
WARDA, IITA, ILCA, ICRISAT, etc.

Writing
Editing
Publishing

Research
Community

Libraries

Peace Corps
Programmes

DONORS
IDRC, UNDP, USAID, BFW...

Extension
Workers

Programme
Managers

Farmers'
Associations,
other NGOs

Figure 1.   Information exchange via *Entre Nous*.

IPM activities of the US Peace Corps

K. BYRD

US Peace Corps, 1990 K St. NW, Washington, DC 20524, United States

The US Peace Corps was created by an act of Congress in 1961. This act was initiated by the vision of President John Kennedy to allow Americans to serve abroad and help improve the quality of life for other people. Although technically a government organization, the Peace Corps has volunteers working at village level, a characteristic normally associated with NGOs. There are approximately 5500 Peace Corps members working in 80 countries.

The Peace Corps' work began in Ghana. The presence of the Peace Corps in any country is at the invitation of the host government. As guests, the staff members collaborate with the host country agencies to address problems in that country. Each country's problems are different, and the role that volunteers play will also differ, so once this has been defined, a plan of action is drawn up and volunteers recruited for the assignment. Areas of assignment usually fall into the sectors of health, education, agriculture, small business development and urban development, but also include the cross-sectorial areas of water and sanitation, youth development and women in development.

Volunteers undergo extension training in a programme designed to provide them with the necessary skills to communicate at the local level, to adapt to cultural differences and to promote appropriate technology. This pre-training varies in duration from 12 to 20 weeks depending on the country and particular programme. After the successful completion of this training programme, the trainee is sworn in as a Peace Corps Volunteer (PCV) for a period of two years. The volunteer is then assigned to a pre-selected village site with housing and a living allowance comparable to his/her counterparts. The placement of the PCV in a village and the subsequent interactions enable the PCV to understand the needs of the villagers better. The PCV can then respond as a resource person in addressing some of the problems affecting the villagers.

Constraints to the deployment of IPM strategies related to the Peace Corps are as follows.

(a) There can only be as many volunteers in each country as agreed by Peace Corps and the host country. This number varies from country to country, and is based on programming that reflects areas of intervention by PCVs.

(b) If host country agencies prefer Peace Corps to work in certain areas only, PCVs may not be trained to address problems in a particular sector.

(c) The Peace Corps can locate volunteers with specialized skills in a certain sector, but not necessarily on a consistent basis within a project to sustain the work that they may need to perform over a period of time.

(d) Support funding is provided by the regional offices that oversee the activity of the PCVs in the countries of that region. This funding is restricted to recruiting, training and supporting the PCV.

At present the Peace Corps, with financial support from USAID, is implementing a programme of activities to address the problem of pesticide misuse and abuse in Central America. These activities will provide training for government extension agents, PCVs, farmers and households in pesticide management and the safe use of pesticides. The Peace Corps will collaborate with local research institutions to provide appropriate technology training for the target population. It is felt that the Peace Corps relationship with the village and farming community, in collaboration with the research institutions and funds from USAID, will bridge the gap between extension and research and provide an effective mechanism for the transfer of skills to address the problems associated with pest control. A similar collaborative effort could work in Africa.

APEMAF's current pest and vector management activities

A.A.R. LATIGO

African Pest and Environment Management Foundation (APEMAF), P.O. Box 14126, Nairobi, Kenya

INTRODUCTION

The African Pest and Environmental Management Foundation's (APEMAF) projects on integrated pest and vector management are designed to develop and strengthen national programmes to achieve sustainable cost-effective crop protection and vector control methods. These must be within the capabilities of the end-users and safe to humans and the environment. The goal is to increase the level and reliability of crop production and to reduce the debilitating effects of vector-borne diseases, especially malaria and bilharzia.

The adoption of IPM has been slow, especially in African countries. This could be attributed to the following constraints.

(a) Inadequate research: IPM requires a sound understanding of the system for which there are limited research workers.

(b) Inadequate extension service: IPM technology is complex and more difficult to transfer than pesticide application technology. It requires effective extension services, which do not generally receive adequate support.

(c) Complexity of IPM technology: as IPM is more complex than conventional pest or vector control adoption is slow because farmers dislike adopting any innovation that is difficult or time consuming to implement.

(d) Inadequate community involvement: communities are often considered passive consumers of technology and as such they have not been involved in the identification, planning, implementation and evaluation of IPM programmes.

APEMAF is aware of all these problems and how they relate to food production, human health and the environment, and has become increasingly involved in IPM programmes.

CURRENT PEST AND VECTOR MANAGEMENT ACTIVITIES

Since its inception in 1986, APEMAF has identified nine projects with pest and vector management activities in Kenya, Uganda and Tanzania. Two current projects have IPM as their primary focus, while the remaining seven incorporate support training and information in pesticide or pest/vector management. One project involves laboratory and field tests and another is designed to support chemical vector control using insecticide-impregnated materials as a component of an integrated malaria control programme.

These projects are variously funded by the Australian Government, United Nations Children's Fund (UNICEF) Regional Office for Eastern and Southern Africa, UNICEF Uganda, Germany (through the German Academic Exchange Services (DAAD) and Deutsche Gesellshaft für Technische Zusammenarbeit (GTZ), and recently, the IPMWG.

These projects are run by APEMAF's Centre for Environmental Risk Assessment and Pesticide Information which has a regional laboratory in Kampala, Uganda. The current programme includes the following.

(a) Systematic assessment of pesticide hazards related to people and the environment; surveillance of environmental trends in the region and provision for early warning of significant changes in the environment.

(b) Pesticide quality analysis and disposal of unwanted pesticides.

(c) Development of environmental guidelines, including those for IPM, to help communities understand mitigation measures for specific environmental and pest/vector management problems.

(d) Individual projects for supportive measures important for IPM implementation and environmental protection. These include:
  • regional database to increase access to environmental and IPM information;
  • postgraduate training in environmental risk assessment and management, including IPM;

- intensive training in technology and information transfer for community extension workers;
- involvement of communities and schools in environmental and IPM information and technology transfer.

APEMAF's pilot project on a 'small-scale community participatory IPM programme' started in 1990. It involves two communities in eastern Uganda with a total population of about 22 000. More recently the project was extended to Kenya and covers a community with a population of 15 000. These projects are based on the fact that the community is very knowledgeable about pest and vector control. APEMAF has involved the communities right from the identification and implementation of the pilot projects to integrating IPM into traditional and indigenous practices.

The participatory community IPM programme aims at:

(a) collecting knowledge about traditional and indigenous methods of crop protection and vector management from the communities;

(b) organizing village-level meetings/seminars for farmers to exchange experiences and methods from different ethnic groups; promising components are selected for evaluation;

(c) establishing a network of trials in and around farmers' plots to evaluate the selected components of IPM.

PROGRESS OF THE IPM PROJECTS

At farmer level the following components have been identified through community participation and are being tested in the pilot projects.

(a) Integrated vector management for the combined malaria and bilharzia projects in Uganda:
   - environmental measures, e.g. vigilance on breeding sites for mosquitoes and snails (intermediate hosts of bilharzia);
   - two common plant species have been identified that reduce man-mosquito contact;
   - cow-dung as a means of deterring mosquitoes;
   - papyrus mats (treated with safe insecticide) for reducing malaria transmissions as an alternative to heavy pesticide application;
   - Endod (a soapberry plant that grows locally) has been identified as a means of control of both mosquito larvae and freshwater snails (as an alternative to synthetic molluscicide).

(b) Integrated pest management in Uganda. The above experience was used to initiate a pest management programme in the same area. The communities have the following to offer to the relatively new project:
   - crop storage methods in fire places, using ash, and in special granaries as a means of crop protection;
   - manipulating the planting seasons for rice as a measure against quelea and other grain-eating birds;
   - selecting a variety of rice for early maturity which is unacceptable to birds;
   - burning vegetation to deter cowpea pests.

These two examples illustrate the benefits of involving the farmers and the community in adopting an IPM programme. The projects encourage the communities to select workers who are trained to follow-up the activities. The communities also contribute to financing certain inputs of the projects which are not available locally, which reduces dependency on external support. The participation of the community is essential to build a good foundation for IPM.

The Research and Technical Applications Group

Y. SANOGO

Groupe de Recherche et d'Applications Techniques (GRAT), B.P. 2502 Bamako, Mali

The Research and Technical Applications Group (Groupe de Recherche et d'Applications Techniques, GRAT) is an NGO set up in Mali in 1980. GRAT is concerned with:

(a) applied research in a number of areas, including appropriate technologies, new and renewable energies, and the environment; and

(b) supporting local communities in, for example, agriculture, processing and packaging of agricultural produce, and commercial forestry.

GRAT co-operates with the national plant protection service as part of its work at community level. This involves receiving technical advice, requesting technical support and training for GRAT field agents and members of the local groups with whom it works. For example, a GRAT organizer and the leaders of a women's group received training, and the group was promoted to a plant protection brigade by the National Plant Protection Service at Mopti (the central delta of the River Niger in Mali).

A co-operative crop protection programme, involving GRAT, AFRICARE and the Peace Corps, has recently been initiated. This provides trainers and literate farmers with literature aimed at encouraging the correct use of pesticides and the promotion of natural plant protection. Non-government organizations such as GRAT are in a position to make a valuable contribution to IPM programmes. However, this requires co-operation at different levels, for example:

(a) between NGOs, research institutes and specialists on the one hand and local communities on the other;

(b) training of NGO field agents by research workers and specialist organizations with the aim of more effectively reaching local populations.

IPM Activities of AFRICARE

D. GERBER
AFRICARE, B.P. 1792 Bamako, Mali

AFRICARE is an American private voluntary organization, established in the early 1970s by African-Americans and Africans, with two main purposes:

(a) to improve the quality of life for rural Africans;

(b) to increase Americans' knowledge of Africa.

Funding for the programmes comes from USAID, foundations, corporations, other bilateral and multilateral donors and many private individuals.

AFRICARE – Mali is one of the oldest and largest AFRICARE programmes. Projects are established in water resources, agriculture, natural resource management, small enterprise development and health. The focus of the programmes is on self-help development. The organization works closely with governments and other development agents, but most importantly with village groups.

The field agents are key personnel who are responsible for listening to the villagers and helping them set up development programmes. One large programme in Mali is vegetable gardening. The project is associated with 24 gardens and each has problems with pests. The agents, like those of many other NGOs, often do not have any knowledge of IPM, but are the key people to deal with the problems of the gardeners. They have in the past sought help from government extension agents who often do not know about the use of pesticides or alternatives.

The focus of a project currently being set up with GRAT and the Peace Corps, in collaboration with the Service National de Protection des Végétaux (SNPV), the Institut d'Économie Rurale (IER) - Section des Recherches Fruitières et Maraîcheres, the Institut du Sahel and DNAFLA (literacy training organ of the government), is to bring information regarding the sale and use of pesticides and the possibility of alternatives to extension agents and farmers, particularly gardeners. A book with information on pests, pesticides and alternative control measures is to be published in French and Bambara. Peace Corps volunteers, NGOs and government agents will receive these manuals and be trained in their use.

Two visual aids are to be developed for use by extension agents and villagers. These will depict stories promoting the safe use of pesticides and the possibility of alternatives and will be an important addition to the manual in spreading information.

In Mali when pests attack gardens, the gardener has neither tools nor a rational strategy to deal with the problem. In desperation, if a product is available it might be used regardless of its potential value or harm. In this instance agents, NGOs or government organizations are of little help.

In a garden in the Sikasso Region the villagers have, in desperation, used RAMBO and Shell-tox insecticide spray as well as insecticide powder (which came in clear plastic bags with no markings) to control pests. Not only are such practices inherently dangerous but they are also ineffective.

So far this workshop has not discussed the role of pesticide manufacturers and insecticide importers in promoting safe use of chemicals and IPM. West African governments should address these problems.

In summary, there are many constraints to IPM implementation:
- extension agents do not know enough about IPM;
- farmers and gardeners do not know enough about IPM;
- pesticides are often not properly used and the appropriate ones are not available;
- there is a lack of liaison between all the agents, including NGOs, working in the country;
- there is a lack of information-sharing among West African countries and agents.

CARE International

J.-M. VIGREUX
ANR Program Co-ordinator, B.P. 143 Maradi, Niger

The development of IPM is hampered by two factors: the difficulty of making the techniques acceptable to farmers and the lack of funds to evaluate the programmes. Those responsible for funding programmes often fail to appreciate the impact of IPM, while both national and international agencies responsible for implementing programmes are sometimes unable to persuade the funding bodies to continue the effort put into developing IPM.

The first obstacle of overcoming farmers' reluctance needs to be tackled by making more use of farmers' own knowledge and expertise in order to persuade them of the potential benefits of IPM. Farmers need to be involved from an early stage in the process of research and development. This involves breaking out of the conventional research-and-development mould as set out in the IPMWG report and, instead, aiming for a more participative research. This involves not only consulting the farmers, but also giving them an opportunity to contribute to the choice of available options. As has been proven with other agricultural research and development, participation is an essential stage for research workers, extension workers and farmers.

Three projects carried out by CARE International in West Africa have involved participative research. All three projects have been highly regarded by both funding bodies and national partners. However, such satisfactory results—achieved so far only on a limited scale—require the approval and backing of the decision-makers, including research workers. This is especially important in view of the corruptive pressure exerted by the importers of plant health products. Because of these constraints, as well as political and institutional constraints, research workers also need to:

(a) direct their research towards the real environment, which is always less comfortable than the controlled environment of the laboratory;

(b) abandon their intellectual self-satisfaction and pay greater attention to the search for solutions which are simple to apply;

(c) become less intellectually remote and listen to the farmers;

(d) reject over-specialization in favour of research which recognizes the multidisciplinary concerns of the farmers.

CARE International

D. MIDDAH
ANR Co-ordinator, B.P. 143, Maradi, Niger

Two fundamental questions need to be asked about the setting up of integrated pest management (IPM) programmes:

(a) does the whole basis of the applicability of IPM need to be re-thought?

(b) should problems relating to the form of IPM be raised? This involves clarifying what is meant by IPM at the national level, as well as at the level of regional organizations, funding agencies and national or regional development organizations, for example, the NGOs.

CARE International—and CARE Niger in particular—has a policy on the use of pesticides which highlights the dangers of the uncontrolled use of plant protection products. At the same time, it encourages the search for alternative approaches by looking at the range of local methods, and testing these in co-operation with farmers.

As part of a project of institutional assistance to agricultural extension services, CARE International is developing a programme of participative research based on extension services in a direct relationship with farmers and research institutions. With the experience already gained in project development, especially the participative approach, it is clear that there are constraints to the satisfactory implementation of these approaches. IPM is not immune to these constraints, some of which are listed below.

(a) The poor definition of project objectives in relation to the activities required by some funding agencies.

(b) A sectorial view of the programmes and the resulting approaches.

(c) Diagnostic approaches which are not satisfactorily carried out and which yield little information.

(d) The training or profile which agents in the field should have (multidisciplinary).

(e) The non-viability of local village organizations, for example, co-operatives in Niger.

(f) The political definition of priorities on development policy, for example, poor pleading of cases by technicians.

Discussion

The discussion following presentations by the panel of NGO representatives centred around issues of pesticide use, participatory forms of research, and the role of NGOs in the implementation of IPM.

G. Schulten commented that problems involving pesticide use and distribution had been raised by several of the speakers. These were important and should be discussed more fully during the course of the workshop. A representative from Togo agreed that pesticide use was a crucial issue, adding that the goal of IPM must be the rational use of pesticides rather than their elimination, and that the problems of out-dated and dangerous substances must be resolved. Y. Sanogo underlined the importance of providing farmers with better information about effective and responsible pesticide use, while A. Youdeowei added that action must also be taken in terms of the development of less harmful substances and an evaluation of the manner in which they are distributed.

A discussion of the role of participatory research in IPM revealed a certain degree of divergence of opinion among various workshop participants. A representative from Benin claimed that J.-M. Vigreux's presentation had been one-sided, and that although participatory methods were useful for certain aspects of research they could not be applied across the board to all areas of agronomic research. He added that it was necessary to clarify research needs first and then to distinguish appropriate forms of research for specific purposes. S. M'Boob argued that one of the main reasons so many IPM projects have failed has been their failure to adopt a participatory approach. In order to improve adoption rates, projects should ensure active farmer participation from the earliest stages.

K. Cardwell asserted that NGOs are well-placed to carry out certain types of research on the ground, such as socio-economic analyses, but that certain aspects of agronomic research, which require a high level of expertise, must be scientifically proven by trained staff before being introduced into the field. She concluded that NGOs had more important and appropriate roles to play at the ground level, than engaging in 'scientific' research.

C. Lomer challenged several of the points made by Dr Cardwell, arguing that farmer involvement is essential at every stage of the research process, and that NGOs have a crucial role to play in providing a direct link 'from the air-conditioned laboratory to the real world'. He remarked that although existing conventions of scientific validation cannot be side-stepped, experience has shown that there is no guarantee that farmers will accept new technologies, however appropriate and well-tested researchers claim them to be. He concluded that both forms of research must run in parallel, and that researchers must involve farmers in their work from the earliest stages for their work to reach fruition.

A representative from Nigeria acknowledged that participatory research was useful to ensure farmers' consent and confidence, but inquired whether any information of value to IPM had been acquired from farmers. A. Latigo responded by pointing out numerous examples of indigenous technologies and practices which have proved effective for pest control and which, with further research, could contribute to the development of new and improved IPM strategies. Such practices include the use of certain species planted around compounds or fields to repel pests, mixing grains and cereals with ash for long-term storage, various forms of intercropping, using smoke to protect stored grains, as well as burning green foliage in fields to repel pests. He also cited the specific example of a local leaf which is soaked in water by certain indigenous groups and used as a cure for malaria.

Finally, several comments were made pertaining to the role of NGOs in the development and implementation of IPM strategies. A representative from Niger began by expressing the opinion that CARE should address a wider spectrum of rural development issues than those of pest management in order to respond effectively to the needs of rural people. S. M'Boob emphasized that unlike many government extension services, NGOs in Africa had the distinct advantage of possessing direct contacts at the grass-roots level. He asserted that one of the goals of the workshop should be to discuss the capacity of NGOs in this respect and to identify ways of strengthening their present role. A. Diop confirmed that, in his opinion, NGOs in Africa had achieved considerable success in reaching and training farmers at the grass-roots level. He added that NGOs were now in a position to develop more effective methods of participatory research and extension, and to promote increased confidence and communication among farming communities.

A representative from Nigeria expressed concern over a number of problems that foreign NGOs could encounter. These included complications in gaining government clearance, difficulty in effectively communicating with farmers due to linguistic and cultural differences, as well as the danger of creating contradictions and confusions for farmers in cases where both NGOs and government extension services are present. A. Latigo said that, international NGOs must obtain government clearance before they are allowed to operate in a country and often work in close collaboration with a government counterpart.

The main conclusions from this discussion period were that the importance of farmer participation in IPM research and implementation cannot be over-emphasized nor can the necessity to utilize mechanisms which have been developed (i.e. by NGOs) to achieve grass-roots participation.

IARC PERSPECTIVES

Rice integrated pest management in West Africa, 'the WARDA strategy'

A.A. SY and E.A. HENDRICKS
West African Rice Development Association (WARDA), 01 B.P. 2551 Bouaké, Côte d'Ivoire

The Integrated Pest Management Task Force is one of eight Task Forces that form a link between the National Agricultural Research Systems (NARS) and WARDA. The main objectives of the IPM Task Force can be summarized as follows:
- research co-ordination/mini-network
- transport/tests of technologies
- dissemination of research information
- target assistance

Twelve of WARDA's 17 member countries confirmed their intention to participate in the first meeting of the Task Force in February 1992. Following a review of the questionnaires and the discussions of the IPM thematic groups on phytopathology/nematology and entomology/weeds, it was agreed that:

- blast, brown spot, rice yellow mottle virus, sheath blight and bacterial leaf blight are the five key diseases that should be targeted for joint action;

- gall midge, stemborer, termites and whiteflies should be targeted as the four key insects pests throughout the region.

Each of these constraints would induce the development of research activity with three approaches:

(a) economic impact and etiology/epidemiology of the key constraints;

(b) relative importance of the IPM components;

(c) integration of IPM components and implementation of the IPM approach.

To avoid research duplication and present coherent and credible projects, the Task Force working groups recommended that member countries should put forward research activities which could be co-ordinated in a multidisciplinary approach and at a sub-regional level.

In April 1992, 15 research submissions on diseases, insects, weeds and nematodes were analysed by the Task Force steering committee. Thirteen of the 15 proposals were selected and have been forwarded to the WARDA Director of Research for funding.

The IPM Task Force has also given the steering committee a mandate to explore:

- West African capabilities in nematology and weed science;

- the compilation of a directory of rice pathologists in West Africa;

- the possibility of founding a West African Society of Pathology and the form this should take;

- the importance and timeliness of publishing a manual on identification and integrated management of rice pests in West Africa.

WARDA IPM Task Force activity

A. YOUDEOWEI
*Director of Training and Communications, West African Rice Development Association (WARDA),
01 B.P. 2551 Bouaké, Côte d'Ivoire*

Two major aspects of WARDA's activities which complement the work of the WARDA IPM Task Force are carried out within the Division of Training and Communications.

A training working group has been established and is responsible for identifying the training needs of rice scientists in West Africa, and assisting WARDA in deciding training policies. WARDA will be running a six-week bilingual (French/English) training course on crop protection in rice, based on the concept of IPM, in July/August 1992. During this course, it is hoped to develop and produce a series of training aids on IPM for the major rice pests and diseases in the form of packages of colour slides with relevant texts. Experts in WARDA's IPM Task Force will act as resource people in the training course.

Another relevant initiative is the documentation and communications programme which aims at widespread and efficient dissemination of up-to-date research and development information to rice scientists in West Africa. WARDA publishes a monthly *Current Contents* at *WARDA* which lists the contents of all the new journals and books received in the library. This publication is distributed to all scientists on WARDA's mailing list with a request form to obtain photocopies of the journal articles and book chapters of their interest free of charge. Literature searches are also made for scientists using extensive computerized agricultural databases. The documentation centre has accumulated a sizeable collection of publications on IPM in rice which is available for the work of the WARDA IPM Task Force.

Discussion

A. Latigo commented that A. Youdeowei's presentation had been of great interest to him, especially given APEMAF's recent involvement in vector management programmes related to rice production. He asked if Professor Youdeowei knew of an existing equivalent of WARDA in eastern or southern Africa. Professor Youdeowei responded that although he had not highlighted vector projects in his presentation, vector management was a subject which had received much emphasis at a recent training course organized by WARDA in Burkina Faso. A vector project, planned in collaboration with the World Health Organization (WHO), should begin before the end of the year. Professor Youdeowei was not aware of an equivalent of WARDA in East Africa, but said that the institute was moving increasingly towards continental responsibility in the area of pest control. A. Diop added that the International Rice Research Institute (IRRI) has collaborated with numerous national agencies and, as far as he knew, the International Centre for Tropical Agriculture (CIAT) was present in East Africa.

Y. Sanogo noted that A.A. Sy had not included mice, rats, fish or mammal pests in his presentation, all of which represent a considerable problem in rice production in Mali. He also questioned why Mali was not included in programmes described by Mr Sy. Mr Sy responded that the IPM projects described in his presentation were of a limited nature, and that although the projects were aware of these pests, specific expertise for their control was lacking. He added that although individual officials in Mali had been contacted, sufficient feed-back was not received to ensure Mali's inclusion in the project.

Removal of constraints on IPM implementation: the perspective of an international research centre

W.N.O. HAMMOND

International Institute for Tropical Agriculture (IITA), B.P. 08-0932 Cotonou, Benin

The International Institute for Tropical Agriculture (IITA) has had a vital role to play, both in the search for solutions to crop protection problems and in their implementation at regional, national and local levels. The accidental introduction of exotic pests in Africa, provides IITA's scientists with a unique opportunity to raise awareness among decision-makers of the existence of alternatives to chemical control, and to train crop protection researchers and practitioners in the techniques necessary to implement these alternatives, especially biological control.

A central part of IITA's programme has been the incorporation of adequate levels of pest and disease resistance into higher-yielding crop varieties. Since many of the concepts and techniques of biological control are unfamiliar to traditionally trained decision-makers, its introduction has involved much more than the technical training of practitioners. Great emphasis has been placed on helping to establish and provide continued support for co-ordinated biological control and ecologically sound pest management groups within national agricultural research and extension services. The involvement of socio-economists and farmers in the identification of pests and the implementation of appropriate management packages has been incorporated in the development of pest management projects.

Constraints and opportunities in implementing IPM in West Africa: perspectives from an international agricultural research centre

O. YOUM

International Crops Research Institute for the Semi-Arid Tropics (ICRISAT), Sahelian Center, B.P. 12404 Niamey, Niger

INTRODUCTION

Opportunities exist for the successful implementation of IPM in West Africa. However, there are a number of constraints which need to be addressed at national, regional and international levels to enhance the impact of IPM in sustainable agricultural production. Constraints such as the occurrence of multiple pests across countries and across commodities, a lack of, or poorly developed infrastructures, a lack of understanding of IPM complexity, principles and concepts, limitations in human resources and trained IPM specialists, lack of a full exploration of farmers practices, perspectives and understanding on the role and impact of IPM, and a lack of, or poor information and documentation services need to addressed before IPM can be successful. International Agricultural Research Centres (IARCs) in partnership with NARS, working with farmers and extension services, regional projects and organizations, have and will continue to play a pivotal role in addressing many of the problems and so enable the farmer to sustain agricultural production.

Successful development and implementation of IPM in West Africa faces many challenges and constraints. Although not exhaustive, the following are some of the major constraints.

AGRICULTURAL RESEARCH AND DEVELOPMENT POLICIES AT THE NATIONAL LEVEL

The national agricultural research and development policy affects the successful development and implementation of IPM in developing countries in many ways. Often agricultural development policies do not consider IPM to be a key component in research and development. Sometimes, an on-going project within NARS provides an excuse for governments to minimize support to existing research stations or national research programmes. NARS should consider the additional funds from donors as a support to strengthening research at the national level, rather than as a substitute for the support that must come from governments.

INFRASTRUCTURE, FUNDING AND DOCUMENTATION SERVICES

Although some NARS have the infrastructures necessary for implementing IPM (often referred to as lead countries in networking terms), many others are less developed in terms of research facilities and operations, in addition to having only limited funding. The lack of long-term funding is often a key constraint to sustainable IPM and agricultural development and production. NARS scientists' management of support projects should be such that the continuous support for research and development at the national level is not inhibited. All too often, donor-funded projects are considered a substitute for the support which should be provided at the national level. The consequences of such an approach is that research activities tend to collapse once the project is phased out or when support funds are stopped. Thus, sustainable funding contributes to sustainable research activities and implementation of IPM. In addition to funding, the development of adequate infrastructures, such as good laboratories with appropriate equipment is paramount. The existence of well-developed information and documentation services, as well as good information management and exchange is important for the implementation of successful IPM.

EXISTENCE OF MULTIPLE PEST PROBLEMS

Although the objectives of IPM are to solve problems caused by pests, the existence of multiple pests on or across commodities increases the difficulties for solving the problems. The management of multiple pest problems requires a multidisciplinary team approach. IARC's approach, such as ICRISAT in partnership with NARS, tackles the problems through close collaboration and a multidisciplinary approach within and across research programmes.

UNDERSTANDING IPM CONCEPTS, PRINCIPLES, PHILOSOPHY AND COMPLEXITY

IPM has many definitions based on different perceptions and this may lead to confusion in the terms and approaches used. IPM is an integration of all compatible techniques for use in pest control. It means more than just pest control, and also includes many other aspects, such crop and land management using a multidisciplinary approach. To implement IPM, farmers need to be involved at an early stage of IPM technology generation, development and implementation. Training on IPM philosophy, concept and principles is important if IPM is to be implemented successfully.

HUMAN RESOURCES DEVELOPMENT AND THE LINKAGE BETWEEN RESEARCH AND EXTENSION

In many instances, NARS have one scientist working on several pests and crops. This constitutes an overwhelming responsibility which in many instances contributes to limited progress. The importance of training is evident and has to be emphasized. Another key constraint is the poor links between research and extension. In this regard, NGOs in collaboration with IARCs and NARS have a key role to play in applying the results of research to the farmer's fields.

COLLABORATION BETWEEN NATIONAL/REGIONAL DONOR-FUNDED IPM PROJECTS

Regional projects, extension services and national agricultural scientists with international research institutions can all collaborate to increase the efficient implementation of IPM. IPM is a long-term ecological approach to pest control. Short-lived IPM projects cannot be justified unless they provide continuity, support for research, training and development. Unless a long-term approach is taken, short-lived and poorly funded IPM projects will be less efficient in solving pest problems.

OPPORTUNITIES FOR IMPLEMENTING IPM IN WEST AFRICA

Suitable conditions exist in many parts of West Africa for the development of IPM. There is still a substantial diversity of traditional practices sustained over many generations. The very fact that these practices are still available demonstrates their sustainability. However, the pressure of population increases in many countries has prompted donors as well as government agencies to encourage the use of inputs, such as chemical fertilizers and pesticides. One key element in the development of IPM is the need to overcome a number of the constraints discussed earlier. Training NARS and the development of strong research and extension services is most important. A permanent linkage with farmers is necessary before they start to implement proposed IPM packages. In many instances, farmers' lack of adoption of IPM technology may be due to mistrust, or a lack of conviction concerning its effectiveness. Therefore, an assessment of the impact of a given technology for the farmer is important. Farmers are ready to try IPM packages which they think can help them achieve sustainable and profitable crop production. Strong links between research, extension and farmers at an early stage of development of IPM technology should provide a good basis for successful IPM implementation. Successful IPM depends on team effort and joint collaboration between IARCs, NARS, regional organizations and NGOs. IARCs in partnership with NARS, working with farmers and extension services, regional projects and organizations, have and will continue to play a major role in overcoming many of the constraints to increase and sustain agricultural production for the benefit of farmers.

As an IARC, ICRISAT has a global mandate. Work on crop improvement and resources management in the semi-arid tropics is done through research, collaboration with, and strengthening of NARS. Activities include research and training, and technology generation and transfer through research networks such as the West and Central African Millet Research Network and the West and Central African Sorghum Research Network based at the ICRISAT Sahelian Center in Niger. ICRISAT has five mandate crops, sorghum, millet, groundnut, chickpea and pigeon pea. Its headquarters are in India and activities are located in Africa, Asia and Latin America, but its impact on research and development is felt worldwide. In West Africa, ICRISAT collaborates with NARS through its Sahelian Center and programmes in Niger, Mali and Nigeria. Research programmes in West Africa include the Pearl Millet Improvement Programme, Resources Management Programme, Groundnut Improvement Programme at the Sahelian Center, and the Sorghum Improvement Programme in Mali and Nigeria. Other collaborating institutes/

universities are also based at the ICRISAT Sahelian Center. Collaboration occurs in West and Central Africa through research networks. Research networks address the constraints limiting the implementation of IPM by providing a basis for collaboration on technology generation and transfer, co-ordinating research to minimize duplication of effort and to reduce costs of research, training, exchange of information etc. ICRISAT also provides strong support for information and documentation services which are often lacking in many national programmes in West Africa. Several ICRISAT databases are available and accessible to NARS. Among the most used is SATCRIS (Semi-Arid Tropical Crops Research Information Service) at the ICRISAT Center in India. International and regional workshops, field days, consultancies and training activities are among the activities organized by ICRISAT to foster effective information exchange and collaboration with NARS.

At the ICRISAT Sahelian Center and in its West African programmes IPM activities have a multidisciplinary approach and focus on pests of millet, sorghum and groundnut. Main activities include the development and use of resistant varieties, improved cropping systems, the identification and use of insect pheromones, the exploration and use of beneficial natural enemies, surveys of farmers' traditional pest management practices, and training. Close collaboration with other institutes has achieved substantial progress in basic research. In collaboration with NRI for example, components of the sex pheromone of the millet borer have now been identified and research results show that three of the five components in an optimum ratio attract as many males as females. Millet stem borer pheromone technology was developed in collaboration with farmers, thus, giving them the opportunity to be involved at a very early stage of technology development. Approaches such as these further IPM, both at the developmental as well as at the implementation levels.

In conclusion, the development of IPM and its implementation in West Africa faces a number of challenges and constraints. Its successful implementation depends on collaboration between IARCs, NARS, NGOs, national and regional projects and institutions, as well as the involvement of farmers at an early stage of development. The role of IARCs such as ICRISAT in addressing constraints and development of IPM for sustainable agricultural production is evident.

Discussion

Following these presentations, the chairman Professor Youdeowei underlined the prime importance of training IPM practitioners, an area in which IITA has developed considerable expertise.

Following the presentation by O. Youm, S. M'Boob commented that although classical biological control could be viewed as one of Africa's few success stories, greater progress could have been made if there had been more farmer involvement at all stages of the programme. For example, the problem of continued pesticide use by farmers during biological control programmes which nullifies the results, could be avoided if farmers were better trained in the nature and functioning of this type of control. He cited the example of China where farmers are trained in the raising and release of natural enemies, adding that although the situation of small farmers in China is obviously different from that of Africa, the Chinese experience offers a valuable model for possible IPM initiatives in Africa.

W. Hammond commented that achieving such levels of farmer participation is not easy and perhaps China can show how it is done. Professor Youdeowei stated that WARDA has put increasing emphasis on farmer involvement, especially by building on the farmers' indigenous knowledge. A representative from Togo added that the first goal must not be to include as many farmers as possible, but to deal effectively with small groups of farmers.

P. Matteson underlined the continued important role of national extension services, adding that extension agents must be trained to change their attitudes towards pesticide use and to increase their awareness of alternative pest management strategies.

K. Cardwell warned that biological control programmes could not be sustained indefinitely on the basis of a few success stories and that research aimed at identifying new methods and initiatives is essential. She added that IITA is currently planning a programme (which it is hoped will receive USAID funding before the end of the year) to identify new technologies, suitable for adoption, in collaboration with farmers and extension services.

DONOR PERSPECTIVES

IPM: the perspective from an international organization

G.G.M. SCHULTEN

FAO Plant Protection Service, Via delle Terme di Caracalla, Rome, Italy

The involvement of FAO in the promotion and implementation of IPM started almost three decades ago. It is guided in these activities by the FAO/UNEP Panel of Experts on IPM. Over the years many activities have been conducted and much has been learned of the constraints and solutions for actual implementation of IPM at farm level.

IPM is the leading strategy in FAO's plant protection activities. It is considered an essential contribution towards achieving sustainable agriculture, and it should be developed and implemented within the context of integrated crop management. In assisting member countries FAO works as much as possible in close collaboration with regional organizations, the international centres, international organizations and NGOs.

At this workshop much has been said on the important role of farmers in the development of IPM strategies and the need for novel training methods to target various groups. However, until now little has been said on the role of pesticides in IPM. There is no doubt that pesticides have a role to play in most IPM strategies. There is increasing worldwide concern about the negative impact pesticides can have on man, his animals and the environment and examples of negative side-effects abound.

FAO has developed a *Code of Conduct on the Distribution and Use of Pesticides*, which was adopted by the FAO Conference in 1985, and an amended version to include Prior Informed Consent in 1989. The Code sets forth responsibilities and establishes voluntary standards of conduct for all involved in the distribution and use of pesticides. It clearly recommends IPM as the preferred pest control strategy. Article 3.8 reads:

> 'Concerted efforts should be made by governments and pesticide industries to develop and promote integrated pest management systems and the use of safe, efficient, cost-effective application methods. Public-sector groups and international organizations should actively support such activities.'

In West Africa and particularly in the Sahel, many activities have been, and continue to be, conducted on the distribution and safe use of pesticides. Impressive results have been achieved in biological control of cassava and mango mealybugs, but IPM development and its implementation in the major food and cash crops has progressed little, if at all.

A start was made in the 1980s to promote and implement IPM through the CILSS/USAID/FAO regional project, which has been frequently mentioned at this workshop. This project created national research capabilities in IPM for food crops. The achievements and potential for follow-up activities were almost annihilated by the desert locust plague, which led to massive chemical interventions with substantial donor support. In the aftermath of the plague, the climate for IPM has become unfavourable in many countries because a strong reliance on pesticides has developed to control the desert locusts.

This workshop is very timely to re-orientate national and donor approaches to pest control. The present constraints to overcoming the above problems are, however, considerable. Farmers have become accustomed to chemical interventions conducted by plant protection services or village brigades. When the plague subsided, chemical interventions were continued to reduce grasshopper populations. In recent years, the interventions have been extended to a range of other pests in sorghum and millet and there are clear indications that pesticide use in vegetables is increasing rapidly.

Plant protection services in many countries have become largely donor-dependent for their activities, which are concentrated on chemical interventions and training of village brigades in pesticide application. Donor support is to a large extent focussed on providing inputs, ways and means for their distribution, and training in their use. In consequence, plant protection activities have become pesticide driven.

However, the most serious consequence of the emergency operations against the desert locust is the practice of providing pesticides and applicators, free of charge, to the millet and sorghum farmers in addition to interventions by the pest control services. This unavoidably promotes the use of pesticides and leads to mis-use on crops such as vegetables, for which they were not made available. Under such conditions development and implementation of IPM is impossible.

The constraints to IPM caused by pesticide subsidies, not to mention the free distribution of pesticides, have been felt in many projects conducted by FAO and other organizations. At the fourteenth meeting of the FAO/UNEP Panel of Experts on IPM, held in Rome in October 1989, these constraints were discussed in detail. It was concluded that pesticide subsidies constitute serious constraints for IPM implementation. The Panel recommended that FAO and UNEP should take steps to encourage governments to assess and revise their policies in regard to these subsidies.

In recent years, the need for sustainable agriculture has become widely recognized by the international community. The present system of free distribution of pesticides, or provision of pesticide services, based on annual donations is unlikely to be sustainable. It is also unlikely that in the foreseeable future pesticide costs will be financed from national budgets.

There is a need for change, and a move towards the adoption of IPM. No further discussion is needed on the fact that such re-orientation and actual implementation will be gradual. Policy changes in relation to pesticide usage are required in many countries. It has to be stressed that pesticide donations *per se* are not necessarily wrong from the point of view of agricultural development promotion. They should be seen in the same light as food aid, fertilizer donations, etc. However, when these commodities are provided, there is usually an agreement between the donor and the recipient country that they are sold and that the revenue be deposited in a special account, which can then be used to finance development activities.

As a first step in the development of IPM it is strongly recommended that instead of free distribution of pesticides, in those countries where the practice has evolved, a system is introduced in which farmers contribute towards the costs of the inputs; except in the case of a plague.

It is recognized that the introduction of such a system is not without difficulties, but much can be learned from countries where farmers actually buy the pesticides, and from programmes such as Global 2000.

IPM Collaborative Research Support Programme

R.C. HEDLUND
United States Agency for International Development (USAID), Washington D.C. 20523–1809, United States

The USAID Bureau for Research and Development, Office of Agriculture, plans to fund a new Collaborative Research Support Programme (CRSP) on IPM. Proposals to carry out collaborative research will be requested from US universities in late spring 1992.

Each proposal must include collaboration with scientists in one or more developing countries and will utilize a technical pest control problem to address constraints to the implementation of IPM. These constraints, identified by a multidisciplinary planning team of senior scientists, are the same as those identified by NRI and by the working groups at the workshop. Countries wishing to collaborate in CRSP must have:

- a commitment to implementation of IPM

- an ongoing IPM research and implementation programme

- a willingness to provide resources to the research effort

- existing institutional and management capabilities in IPM

- a willingness on the part of the USAID offices in the country to support a research effort in IPM.

USAID commentary on pest management in West Africa

W.I. KNAUSENBERGER

United States Agency for International Development (USAID), Washington, D.C. 20523–1809, United States

Since 1975 it has been explicit USAID policy to promote IPM as the primary strategy of crop protection within its agricultural development programmes. USAID has been supporting crop pest management research and development largely through commodity/systems-orientated research. Few specific pest managment projects have been initiated in Africa by USAID since the USAID-funded CILSS-FAO project on the research and development of IPM of basic food crops of the Sahel, 1978–87.

This ambitious project focussed on establishing an IPM research capability in the Sahel, and developing IPM strategies for major crops. It was intended to complement USAID's Regional Food Crop Protection Project, 1975–85, which was primarily an institution-building project to help participating CILSS countries establish or strengthen national crop protection services through training, construction of facilities, and provision of equipment and supplies.

Some people feel these projects have been failed because they did not achieve their ambitious objectives. Yet they were not so much condemnations of IPM as a discipline and strategy, as of an unwieldy project over too large a region, with the associated management problems of establishing liaison at research, development and farmer levels. Despite many problems the two Sahelian pest management projects succeeded in creating functional national crop protection services and a regional IPM research capability. Crop protection services had hardly existed before USAID trained and equipped many of them. As a result of this experience and in agreement with the mandate in USAID to 'focus and concentrate' and to 'bilateralize' projects, there is a distinct move away from large regional projects.

Currently, the only other regional pest management organizations in the Sahel are:

- the DFPV in Niamey, an autonomous unit within CILSS, and administratively linked to AGRHYMET;

- the Organization for Control of Grasshoppers, Locusts and Grain-eating Birds (OCLALAV) which is based in Dakar, and focusses on regional monitoring and information exchange regarding migratory pests.

Several donors have supported crop protection strengthening activities in the Sahel on a bilateral basis, notably:
Canada (CIDA);
INSAH, Niger, Burkina Faso;
France (CIRAD, ORSTOM)—Sahel-wide, CIRAD especially with respect to grasshopper and locust control;
Germany (GTZ)—Niger, Mauritania, Mali-proposed;
The Netherlands (DGIS)—Chad via FAO, Niger-DFPV, Senegal via FAO;
the United Kingdom (Overseas Development Administration (ODA)-NRI)—Mali;
USAID—Guinea Bissau (to 1990), Mali and Niger;
the EC (Technical Centre for Agriculture and Rural Co-operation (CTA)) INSAH, and
UNDP through FAO—Chad, Mali and Senegal.
Other donors have been involved, particularly during the locust/grasshopper campaigns of 1986–90.

A number of international agricultural research centres are active in the region and many of these have pest management research components.The work of the IITA Biological Control Programme in Benin is especially relevant and it liaises with the DFPV at AGRHYMET. However, co-ordination of their activities, those of the NARS, and of the bilaterally funded pest management activities is not taking place. INSAH's UCTR/PV has a clear role to play in promoting such co-ordination.

INSAH'S PEST MANAGEMENT FUNCTIONS
Since the end of the USAID/CILSS/FAO IPM project in 1987 INSAH, recognizing the importance of IPM to agriculture in the Sahel, has maintained information exchange and co-ordinated the nine

National Crop Protection Services and NARS. INSAH created a Pest Management Technical Co-ordinating Unit, UCTR/PV, specifically to take on this role. It is a programme in the plant production section of the Department of Research on the Environment and Agriculture. Through this unit, INSAH aims to co-ordinate pest management in the Sahel with the IARCs, NARS, national crop protection services and other regional and international organizations. The only other organization doing similar work in Africa is PESTNET of ICIPE in Nairobi, Kenya.

The UCTR/PV unit has functioned very effectively with a small number of staff: the scientific co-ordinator (Dr Ba Diallo Daoulé), two secretaries and a driver. Limited funding has been received from CIDA, UNDP, CTA and others for co-ordination and information dissemination activities. The budget for this unit has been drawn from *ad hoc* projects, not core funds. The productivity of this unit is not sustainable, and its potential cannot be realized under the current staffing pattern.

UNSO has committed itself to providing resources to UCTR/PV for an entomologist (2.5 years) to focus on pest population surveillance, crop-loss assessment and food security issues as related to pest management.

FIVE-YEAR PLAN OF INSAH—UCTR/PV

INSAH's five-year plan (1990–94) in crop protection has four aspects, as elaborated with Sahelian crop protection researchers in a number of fora organized by UCTR/PV.

Sahelian Network for Research on IPM of Crop Pests

Eleven priority research themes or targets were identified by Sahelian crop protection researchers at INSAH's second conference on food crop IPM in the Sahel (Bamako, January 1990). These are:
- parasitic plants (e.g. *Striga*) and weeds
- stored product pests in rural areas
- vegetable crop pests
- rice pests
- insect pests and diseases of millet, maize and sorghum
- rodents injurious to food crops, tree nurseries and young agroforestry plantations
- indigenous plants with natural pesticide properties
- grain-eating birds
- grasshoppers and locusts
- ecotoxicology and environmental impact of pesticides.

Each of these areas has been more fully defined and described in terms of the countries likely to be involved; possible or actual executing agencies; status of research in the Sahel; justification; objectives; proposed IPM research programme strategies; pilot demonstrations at farmer level; training plans and expected results.

INSAH-UCTR/PV proposes to promote and track these aspects in each of the priority topic areas, and play a role in identifying sources of support for various aspects of the research.

Strengthening of crop protection services

The following are the proposed elements for activities in this area:
- participation in the development of co-ordinated training plans for national crop protection services
- promotion of training at the farmer level
- promotion of regional plant quarantine measures and regulatory structures
- promotion of the development of a regional laboratory for pesticide quality control and pesticide residue testing.

Information dissemination

The following activities are proposed:

- continued publication of *Sahel PV Info*, which has appeared 10 times annually since April 1988; it provides a useful service and forum, publishing information on current phytosanitary developments, research results, relevant workshops and conferences, new publications

- publication of the proceedings of annual meetings of Sahelian crop protection research workers

- preparation of brochures and fact sheets derived from the US AID-CILSS-FAO IPM project results

- exploration of the possibility of launching a scientific review journal in Sahelian crop protection, with two issues per year.

Funding for these publishing, development and dissemination activities has been supported mainly by CIDA and CTA.

Regional co-ordination

The final aspect of the five-year plan is to promote regional co-ordination by providing regular fora for discussion and information exchange, specifically by organizing:

- annual meetings of crop protection researchers, national crop protection service directors, and 'chefs de base' of the main plant protection field stations

- technical conferences relating to IPM of food crops in the Sahel

- *ad hoc* meetings addressing specific topics or needs.

Discussion

Following the presentations by G. Schulten and R. Hedlund, a representative from IITA underlined the crucial role played by FAO in terms of funding, logistical support, networking, co-ordination, research and policy in IPM, and expressed the hope that the FAO would continue its essential contributions to the development of IPM in Africa.

R. Hedlund was asked to elaborate on CRSP procedures and clarified that while only US institutions were eligible to apply for funding through the global research programme he had described, one of the criteria for selection was collaboration with a USAID-assisted institution in the South. He therefore encouraged representatives from the South with an interest in the programme to approach US institutions. A. Latigo expressed the hope that Mr Hedlund could provide assistance in initiating and securing such contacts between North and South. A representative from Nigeria, who was concerned with the possibility of contradictions between the different branches of CRSP, was reassured by Mr Hedlund that this was unlikely since each CRSP programme would include an IPM component.

K. Cardwell expressed regret that more donors, eg. CIDA, African Development Bank etc, were not present at the workshop and suggested that once the proceedings were published collaboration with these additional donors be followed-up. M. Iles responded that he would try to ensure the continued participation of donors, both those present at the workshop and those who would be present at the meeting in Istanbul in May 1992.

T. Wood, representing NRI, noted a recent preference on the part of donors to provide funds for the implementation, rather than the development, of IPM technologies, since numerous existing technologies have not been adopted.

F. Vicariot concluded the discussion by noting that donors are not always attentive to the needs of their beneficiaries, often funding projects based on their own agendas. He emphasized that perhaps the most important aspect of this workshop was to allow dialogue between donors and recipients, and to provide in writing the priorities and proposals of recipients for consideration by donors.

COUNTRY PAPERS

Integrated pest management in Benin

C. LAWANI
Service Protection de Végétaux, Direction de l'Agriculture, B.P. 58 Port Novo, Benin
and
A. KATARY
Direction de la Recherche Agricole, B.P. 715 Cotonou, Benin

INTRODUCTION

Both the government and IITA are carrying out biological control in Benin. The main aim is to protect the environment by combining all forms of pest control in a more efficient way. A mixture of chemical, cultural and biological control methods and techniques are used. These are both cheaper for the farmers, and provide more protection for the environment.

Up to now, farmers have had little involvement in the development of control methods and their popularization. Government departments are concerned with introducing techniques to the farmers, as a preliminary means of popularization.

COMMON PESTS IN BENIN

Benin has, in general, a relatively healthy agriculture, despite the increase in the area over which *Prostephanus truncatus* and *Rastrococcus invadens* are distributed (Figures 1 and 2). However, following abundant early rain, 1991 saw a sudden resurgence of locusts in the north of the country.

Some of the pests which cause economic damage to crops have been the target of biological control applied with the co-operation of the Africa-wide IITA Biological Control Programme. Here, we describe the main pests and the action taken to control them.

Mango mealybug (*Rastrococcus invadens*)

This prolific pest is found on mango trees and some flower-bearing and forest species. Biological control began in 1988, when all species suffered major mealybug infestation. The result was a sudden fall in fruit production, as well as adverse effects on human health and the environment.

Since 1988, follow-up operations and increased releases have continued to maintain distribution levels of the beetles' natural enemy at an appreciable level. This has been successful in reducing damage to below the economic threshold.

Rastrococcus invadens does not go beyond Ina (Bimbérèkè) in the northeast and Alédjo and Sèmèrè in the northwest. Ouinhi, Savè and Ouessè are the main centres of infestation in the centre of the country. On the whole, however, wherever the pest has been identified it is associated with its natural enemy, *Gyranusoidea tebygi*. This seems to indicate that the natural enemy is both better established and distributed.

To supplement the effect of *G. tebygi*, 3350 individuals of *Anagyrus manguicola* have been released on an experimental basis in three localities. The first follow-ups indicate a net fall in the population of the pest, without, however, establishing whether *A. manguicola* had become established or dispersed. Studies are in progress as part of the Biological Control Programme, to investigate this natural enemy further.

Despite the size of the releases, however, some centres of infestation persist—in particular, Pobè (southeast), Sèhouè (south), Ouaké (northwest) and Parakou (northeast).

Cassava green mite (*Mononychellus tanajoa*)

Releases of exotic phytoseiids have been made in several places against this pest of cassava. Observations suggest that:

(a) the population density of the green mite was reduced immediately after release and this fall persisted throughout the growing season;

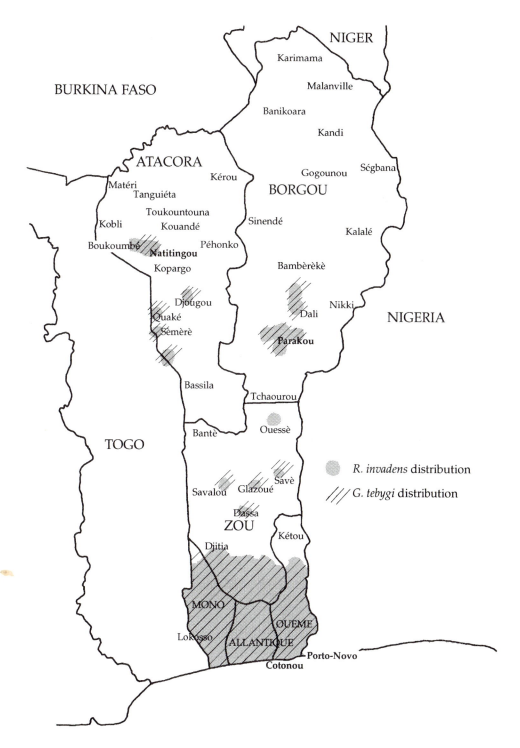

Figure 1.   Distribution of *R. invadens* and *G. tebygi* in Benin, February–March 1991.

(b) no exotic phytoseiids have yet been observed to establish themselves; the various samples made during follow-up showed that only the local phytoseiids, *Typhlodromalus saltus* and *Euseius fustis*, were present.

Larger grain borer (*Prostephanus truncatus*)

As a prelude to biological control, a national survey of the distribution of *P. truncatus* using pheromone traps revealed that this pest was distributed over the entire southern part of the country, while isolated foci were found in the northeast (Figure 2). In the northwest it is mainly the central regions of Togo that are infested (Bassila, Ouaké and Copargo).

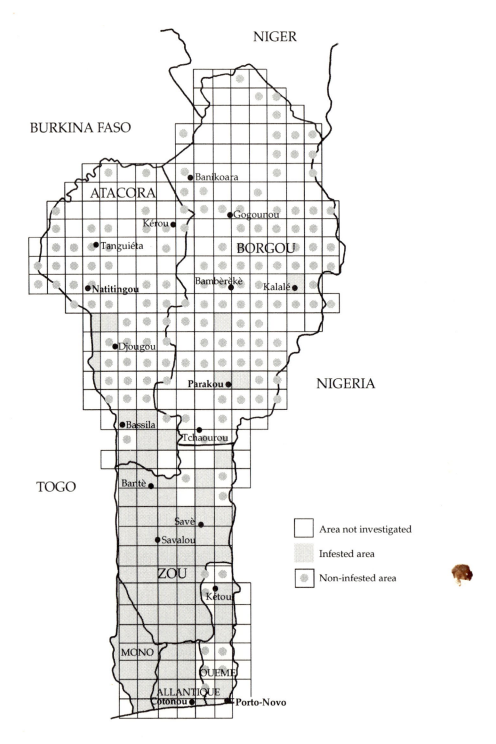

NIGER

BURKINA FASO

ATACORA

● Banikoara

Kérou ●

● Gogounou

● Tanguiéta

BORGOU

Bambèrèkè ●

● Natitingou

Kalalé ●

● Djougou

Parakou ●

NIGERIA

● Bassila

Tchaourou

TOGO

Bantè ●

Savè ●

● Savalou

ZOU

Kétou

MONO

OUEME

ATLANTIQUE

Cotonou ●   ● Porto-Novo

	Area not investigated
	Infested area
	Non-infested area

Figure 2.   Distribution of *Prostephanus truncatus* in Benin in 1991.

The number of *P. truncatus* captured by the pheromone traps varies between 1 and 114 individuals.

Water hyacinth (*Eichhornia crassipes*)

Neochetina eichhorniae was introduced by the Biological Control Programme and was raised in the glasshouses of IITA's station in Benin. Experimental releases were made, and follow-up investigations showed that the natural enemy became established and dispersed in the field.

Other examples of successful biological control include the development of progressive treatment for cotton trees, the release of *Gyranusoidea tebigii* against mango mealybugs and the release of parasitoids against the cassava mealybug.

CONSTRAINTS ON BIOLOGICAL CONTROL

Biological control methods in Benin are constrained by a number of factors. For example, efforts to control the mango mealy bug were not helped by the fact that:

(a) trees were pruned, despite being heavily infested and in spite of farmers being advised not to cut or burn dead leaves in order to encourage the effect of the natural enemies released;

(b) there was systematic use of chemicals, thus destroying released natural enemies;

(c) many of the farmers were disinterested in biological methods because of the length of time these need to become effective; often, this means that they turn to other methods which hinder the establishment and spread of natural enemies;

(d) biological control aims to re-establish the ecological equilibrium between a pest and its natural enemies; this concept is not always understood by farmers, who expect a pest to be eradicated completely.

As part of its future programmes the Plant Protection Department intends to carry out the following:

on cabbage:
- use a parasitoid (*Apanteles*) to control cabbage moth (*Plutella xylostella*)
- use neem to control pests (greenfly, grasshoppers etc.) in nurseries.

on maize:
- release *Teretriosoma nigrescens* to control *Prostephanus truncatus* in the field
- use a binary system for stored maize to control *Prostephanus truncatus*.

Integrated pest management in Ghana

G.A. DIXON, J.V.K. AFUN, H. BRAIMAH, A.R. CUDJOE and A.F.K. KISSIEDU

Plant Protection and Regulatory Service, Ministry of Agriculture, P.O. Box M37 Accra, Ghana

IPM is a term used to describe a system where two or more compatible control measures are directed at a pest to keep its population below economic injury level. It is environmentally friendly and requires a minimum of inputs by the farmers, and so is suitable for a resource-poor farming system, as in Ghana.

In Ghana IPM is not organized at a national level and it is the prerogative of farmers and individual plant protectionists to develop and practise their own pest management programmes. A farmer's first reaction to pest damage is determined by his socio-economic circumstances. Where land is not limited he may increase the area under cultivation and use crop cultivars or land races that are tolerant to the pest. Some farmers adopt farm hygiene practices, such as rogueing of diseased plants, and stubble and crop residues management, to break the cycles of insect pests or pathogens and reduce their latent populations. Farmers may also practise land or crop rotations or fallowing and plant trap crops to reduce the pest load on the crop. To a small extent, some farmers use chemical pesticides alongside these measures. The restriction on entry of exotic pests into Ghana, and the spread of endemic pests within the country, is the responsibility of the government who puts in place quarantine legislations and enforcement through the Plant Protection and Regulatory Services of the Ministry of Agriculture.

There has been little research on the development of IPM in Ghana. Effective individual control methods have been developed for various pests but these have not been incorporated into workable packages. One pest control programme which contained elements of IPM strategy was the control of the cocoa swollen shoot disease (between 1940–60), caused by a virus which is transmitted by the mealybug, *Planococcoides njalensis*. In the programme diseased trees were cut out to remove the virus inoculum and new trees planted. Insecticides were then applied to control formicid ants that attend and help disperse the mealybug, as a means of controlling the pest. Classical biological control of the mealybug using imported and introduced natural enemies was attempted by the West Africa Cocoa Research Institute, now the Cocoa Research Institute of Ghana, without success.

Other institutions involved in pest control activities in Ghana are the Crops Research Institute, Oil Palm Research Institute and Institute of Aquatic Biology, all part of the Council for Scientific and Industrial Research. The Universities, Ghana Atomic Energy Commission, the Environmental Protection Council and the Cocoa Services Division of Cocobod also have an interest in pest control activities.

This paper outlines the present status of pest control in Ghana and the future research, development and implementation of IPM.

INITIATION OF IPM IN GHANA

There is no documented established IPM project in Ghana. The idea has mostly been promoted through relations with international agencies, such as IITA and FAO. The original source of the idea is difficult to trace due to lack of reliable records.

PROBLEMS OF IPM RESEARCH AND DEVELOPMENT IN GHANA

The need for the development and implementation of IPM packages have long been recognized by research workers, especially entomologists. However, their ideas and plans have always remained on the drawing board because of a few basic perennial problems. The most important of these are:

- lack of a national policy and strategy for IPM research and implementation

- poorly funded pest control research and implementation

- inadequate skilled personnel at both technical and scientific levels

- large numbers and complexes of pests

- lack of co-ordination between the institutions involved in plant protection activities

- communication and extension services are weak and farmers are usually too poor to contribute to the implementation of initiatives.

Until recently, agricultural research in Ghana was not well co-ordinated. The pest problems addressed were those perceived by individual researchers, plant protectionists or their institutions, and not those of farmers or of national interest. Now, however, this is changing, since the establishment of the National Agricultural Research Project (NARP) to develop multidisciplinary research and new approaches to technology development and transfer. In the new approach, farmers, extensionists, policy-makers and researchers identify and give priority to problems at a common forum. The research is then directed to provide practical solutions to the problems.

Agricultural research is still poorly funded. This is because there is no fixed national policy and so little financial support is committed to research. The meagre finances available are released in a haphazard and erratic manner. This not only frustrates planning but delays the implementation of projects. As a result, research efforts that could lead to the development of IPM packages are not sustained to a conclusion.

In Ghana, the biggest constraint to the adoption of IPM strategies is the view that yields suffer in comparison with those obtained using chemical pest control measures. The single most important constraint to the development of IPM is the general perception of a constraint on agricultural production. Pest and disease problems are usually ignored during the initial research and development stage until crises arise to disorganize the recommended protection packages. In this situation the crop scientist is under pressure to control the pest quickly. There is no time for biological and ecological studies of the pest problem to determine the most suitable and sustainable control measures. In addition environmental conditions suitable for pest survival and development all-the-year-round cause outbreaks to occur more often. Under these conditions and without time to study the bio-ecologies of the pests, crop scientists resort to *ad hoc* measures, particularly those involving a high dependence on chemical pesticides. Initial encouraging results, lead to the continued use of pesticides with no regard for their immediate or long-term effect on the environment. This view of the role of crop scientists means that low priority is given to their training. This causes a shortage of crop scientists in the country which helps to delay the research and development of IPM.

PRESENT STATUS OF PEST CONTROL IN GHANA

The traditional pest control methods that relied on good crop hygiene and husbandry are still practised. In recent years agronomic studies have highlighted a number of effective methods of control. Ploughing, for example, destroys some soil-borne pests, and manipulation of planting dates enables the most vulnerable stage of some crops to evade peak pest populations. The use of clean planting materials, as well as the depth of planting, soil fertility and water management, enhance crop vigour and enable crops to tolerate levels of pest infestation which would otherwise cause serious damage. Timely harvests reduce field-to-store pest infestations, while stubble and residue management, by burying or burning, kills pests and reduces the carryover populations. The production of commercial crops such as cocoa, cotton and vegetables, involves considerable pesticide use. Setting aside the environmental and health hazards associated with pesticide use, cost alone makes pesticides unattractive. Despite these problems, it is impossible to stop the use of pesticides completely if crop production is to continue to increase to meet the demands of a growing population.

A series of studies have been made on the judicious use of pesticides, especially insecticides. In these a number of chemicals have been screened for specific crops and pests. Economic dosage and application schedules have been determined and recommended to reduce the amount of active ingredients sprayed onto the crop and environment (Owusu-Akyaw and Afun, 1990; Braimah and Timbilla, 1991). In addition, action threshold studies were conducted to help users apply pesticides only when pest populations warrant it, i.e. monitored spraying instead of calendar spraying (Afun *et al.*, 1991). Scouting tracks have been designed for quick assessment of pest infestation or damage, especially in cowpeas (Afun and Owusu-Akyaw, 1991).

Outstanding success has been achieved in the biological control of the cassava mealybug (*Phenacoccus manihoti*), and the mango mealybug (*Rastrococcus invadens*) (Korang-Amaoko *et al.*, 1987; PPRS, 1989). Most of the cited success in IPM components come about as a result of international or bilateral collaborative programmes. The incorporation of resistance in the host is one of the main approaches for future pest control programmes (Atuahene *et al.*, 1992). Research has also identified some plant products, such as wood ash, pepper and ordinary smoke which are effective for the control of stored product pests (Owusu-Akyaw, 1991; Owusu-Akyaw *et al.*, 1992).

All these control methods, most of which are environmentally friendly, have been developed by separate institutions or individual scientists or farmers, and have limited action.

Generally active farmer involvement in design and testing of plant protection activities is very limited if not entirely absent due to the specialized nature of these initiatives.

FUTURE OF PEST CONTROL IN GHANA

IPM is now a serious consideration in Ghana because of the current awareness of adverse effects on health and deterioration of the environment as a result of over-use of chemical pesticides. Chemical usage has left in its wake the following problems:

(a) the development of resistance by target pests which has led to increased numbers and doses of chemicals – pesticide manufacturers respond by developing and exporting more potent and persistent versions, which are used against the stubborn pests;

(b) the development of secondary pests because of the elimination of their natural enemies through the indiscriminate use of pesticides;

(c) the elimination of beneficial non-target organisms, such as pollinators and scavengers;

(d) the contamination of water resources and general pollution of the environment;

(e) health problems for pesticide users and consumers of the produce; these arise from non-adherence to safety precautions and high levels of pesticide residues from misuse of chemicals. In view of this and the need for a sustainable pest control strategy, the government is becoming increasingly committed to IPM.

The decision to adopt IPM strategies for pest problems in Ghana will be one step in the right direction. The development of IPM will be closely linked to the availability of funds, information on the insect/crop relationship and on national agricultural and health policies. Emphasis will be placed on the following:

- cereal (rice, sorghum, millet and maize) pests and diseases, especially stem borers and storage weevils, in particular, the newly introduced larger grain borer (*Prostephanus truncatus*);

- legume (cowpea, soyabean and groundnut) pests and diseases, particular attention will be paid to legume flower bud thrips, pod borers and pod sucking bugs;

- vegetable (eggplant, cabbage, tomato and okra) pests, especially the lepidopterous pests and the leaf feeding beetles;

- tree crop (oil palm, cocoa, coffee, shea nut and cola) pests – priority will be given to the cocoa mealybug and capsids, coffee berry borers and the oil palm leaf miner beetle;

- fruit (mango, citrus, pineapple) pests, with priority given to the mango mealybug, the mango seed weevil, fruit piercing moths and pineapple mealybug;

- cassava and plantain pests and diseases – these include cassava green spider mite, the variegated grasshopper, the plantain or banana weevil, black sigatoka disease and other important pests;

- important environmental pests, such as water weed in freshwater bodies and vectors of animal and human diseases, such as the black fly, mosquito and tsetse fly will be accorded the importance they deserve.

The degree and complexity of pest problems in Ghana are serious, but the human and other resources committed to the problem are small. We advocate a re-orientation in policy towards pest control research and implementation, to underscore the importance of IPM as a tool for sustainable agricultural development and a cleaner environment.

REFERENCES

AFUN, J.V.K., JACKAI, L.E.N. and HODGSON, C.J. (1991) Calendar and monitored insecticide application for the control of cowpea pests. *Journal of Crop Protection*, **10**: 363–370.

AFUN, J.V.K. and OWUSU-AKYAW, M. (1991) Scouting methodology for the management of cowpea insect pests. *Ninth Meeting and Scientific Conference of the African Association of Insect Scientists, Accra, September 1991.*

ATUAHENE-AMANKWA, G., ASAFU-ADJEI, B., HOSSAIN, M.A. and AFUN, J.V.K. (1992) Cowpea varietal improvement activities of GGDP in 1991. *Twelfth National Maize and Legumes Workshop, Kumasi, March 1992.*

BRAIMAH, H. and TIMBILLA, J.A. (1991) The effect of spray arrangement of recommended insecticides on their field efficacy in cowpea pest control. *Eleventh National Maize and Cowpea Workshop, Kumasi, March 1991.*

KORANG-AMAOKO, S., CUDJOE, R.A. and ADJAKLOE, R.K. (1987) Biological control of cassava pests in Ghana. *Insect Science and its Application*, **3**: 905-908.

OWUSU-AKYAW, M. (1991) Evaluation of plant products for the control of maize and cowpea storage insect pests. *Joint SAFGRAD Network Workshop, Niamey, March 1991.*

OWUSU-AKYAW, M. and AFUN, J.V.K. (1990) Maize and cowpea insect pest control. The GGDP experience and recommendations. *Tenth National Maize and Cowpea Workshop, Kumasi, March 1990.*

OWUSU-AKYAW, M., AFUN, J.V.K. and ASIEDU, E.A. (1992) Persistence of some plant products on treated maize and cowpea seeds. *Twelfth National Maize and Legumes Workshop, Kumasi, March 1992.*

PLANT PROTECTION and REGULATORY SERVICES (PPRS) (1989) *Biological Control of Cassava and Mango Pests in Ghana.* Technical Memo.

Integrated pest management in Togo

D. AGOUNKE

Director, National Crop Protection Service, Togo

Until the beginning of the 1980s, chemical control was used throughout Togo. Biological control began in 1982–84 following the accidental introduction of the cassava mealybug *Phenacoccus manihoti* and the complex of cassava green mites. The IPM programme was developed following co-operation between the Ministry of Rural Development and several foreign institutions, for example, GTZ, IIBC, Institut National de Recherche Agronomique (INRA), IITA and CPI-OVA.

This vast programme of biological control gave national protection departments throughout Africa a considerable boost in their search for alternative methods of pest control.

In 1981, shortly after the introduction of the mango mealybug, the larger grain borer, *Prostephanus truncatus*, was discovered in Tanzania and in southern Togo in 1984. Research findings suggested a number of ways of approaching this storage pest, in particular, chemical control using a binary system, as well as biological control. The latter used *Teretriosoma nigrescens*, which had recently been discovered in Togo. The results of all these interventions are currently being evaluated.

Between 1986 and 1987 Togo also saw the accidental introduction of the mango mealybug, *Rastrococcus invadens*, which soon became a major pest throughout Africa. The introduction of *Gyranosidea tebygi* as a means of biological control was not without criticism. However, the experience gained may have been an important factor in changing attitudes towards what could become a national initiative aimed at controlling pests.

More recently (1989–91) Togo has been faced with the problem of the citrus whitefly or aleurode (*Aleurothrixus floccosus*). This might also be a candidate for biological control, such as was practised in the Mediterranean basin between 1970 and 1974.

As far as the cabbage moth, *Plutella xylostella*, is concerned, the approach generally adopted in Togo involves a combination of neem extracts, microbiological control and the enrichment of useful fauna.

One of the main obstacles to the successful application of biological control methods is the lack of reliable quantitative data. Although the pressures of a particular infestation may be felt by individual farmers, and a record may be drawn up by entomologists, quantitative data are not always available. Another factor which is often omitted in evaluating pest control measures is the socio-economic dimension. This was particularly important in the case of measures taken against the mango mealybug where, as with other IPM programmes, the essential objective was to persuade the farmers to adopt the various measures.

DISSEMINATION

Although it is relatively easy to distribute new seeds, resistant or tolerant varieties and effective chemicals or pesticides to farmers, IPM sometimes gives rise to a number of obstacles, in particular, lack of awareness among producers, and a lack of communication between research workers, extension workers and the media.

Conflict of interest between aid institutions or those involved with producers sometimes creates confusion in the field.

IPM programmes now being introduced in Togo pay great attention to the problems, the cost of pesticides, their uncontrolled distribution and use, which may have adverse effects on the health of users, consumers and the environment.

While statistics appear to show that, in general, developing countries use few pesticides, there is no systematic collection of data on cases of chronic or acute poisoning. In addition to cash crops, in recent years there has been an increase in the use of dangerous products on market garden crops in Africa. This has to be seen against a background of absence of any legislation on plant health in some countries, as well as the fact that pesticides are often donated, free of charge, by chemical companies.

Discussion

Following the presentation of country reports from Benin, Ghana and Togo an open discussion ensued which focussed on pesticide use, the importance of farmer involvement and the need for greater regional co-operation and co-ordination. S. Sagnia questioned to what extent successful experiences with IPM, through classical biological control, for example, succeeded in changing attitudes towards pesticide use. The representative from Benin replied that while effective pest control through alternative means will lead to a reduction in pesticide use by farmers, training in the responsible and reduced use of pesticides should be a component of any IPM package. Responding to a question concerning farmer involvement in IPM from P. Matteson, the representative from Benin explained that the IPM project to control cassava mealybug had been initiated as a direct response to the problems of local farmers. It was the farmers themselves who had informed the plant protection services of the cassava mealybug problem.

In response to a question from A. Latigo concerning the use of neem extracts, the representative from Togo said that considerable success had been achieved in IPM through the use of neem, but that its effectiveness in storage was not proven.

Finally, the importance of basing research initiatives on farmers' own expressed priorities was emphasized, together with the need for greater networking and exchange of initiatives, both successful and unsuccessful, at the regional level.

Integrated pest management in Guinea-Bissau

M. CASSAMA and L. ABREU
Ministerio de Desenvolimento Rural e Pescas, P.O. Box 71 Bissau, Guinea-Bissau

The Republic of Guinea-Bissau is one of the smallest countries on the West African Coast, with a land area of 36 125 km². This includes the Bijagos Archipelagos and other small islands off the coast. It borders Senegal to the north and Guinea-Conakry to the south and east. It has a population of around 1 million, which is increasing at a rate of 2.5% every year. About 80% of the population are farmers or craftsmen who live in rural areas scattered in 3600 villages called tabancas. The country is divided into eight administrative regions, which are sub-divided into 37 sectors and the autonomous sector of Bissau City, which has 13% of the total population.

The economy is based mostly on agriculture, which represents nearly 30% of the GDP. About 80% of the total rural population depends on agriculture for its subsistence. Crop production accounts for nearly 70% of the total agricultural production; livestock 20% and forestry 10%. The staple crops are rice, maize, pearl millet, sorghum, cassava, groundnut, sweet potatoes, beans, other vegetables and tropical fruits. Agriculture is the most important source for earning foreign currency.

The family farm sector supplies the greatest part of food produced. There are about 100 000 family farms in the tabancas. Farm sizes do not exceed 3.5 ha. The land is usually divided into different plots providing a complex cropping model which is influenced by both tribal and religious tradition. Modern agricultural inputs, such as fertilizer and improved seed varieties are used for a very small number of subsistence farmers, and few of them receive assistance from extension services. Crops in Guinea-Bissau are susceptible to several pests and diseases which jeopardize agricultural production. Carefully chosen methods of control are essential factors in modern crop cultivation.

In Guinean agriculture the lack of technology, pest problems, ecological factors and socio-economic aspects are among the many causes of low yields and poor quality crops. It is difficult to solve them separately as in most cases they overlap.

Crop damage caused by pests amounts to about a 20–25% reduction in yield. Insect pests can cause serious loss of yield but the extent of crop damage and loss is variable. Following an extensive grasshopper invasion in 1977, the Government of Guinea-Bissau created a National Crop Protection Service, and procured US assistance in organizing and training entomologists, plant pathologists and technicians to apply pesticides. National surveys of food crop pests were initiated and programmes of adaptive research have begun, such as the introduction of predators and parasites against the cassava mealybug.

In Guinea-Bissau, 30 insects and 15 diseases of crops have been recorded. Among them, the following insect pests and diseases are considered the most important:

Insects	Diseases
Heteronychus oryzae	*Pyricularia oryzae*
Spodoptera exempta	*Helminthosporium oryzae*
Zonocerus variegatus	*Sclerospora graninicola*
Coniesta ignefusalis	*Tolyposporium pernicillariae*
Heliocheilus albipunctella	*Cercospora sorghi*
Psalydolytta fusca	*Fusarium oxysporium*
Megalurothrips sjostedti	*Alternaria solani*
Plutella xylostella	
Analeptes trifasciata	
Helicoverpa armigera	
Maliarpha separatella	

No systematic integrated approaches have been initiated in Guinea-Bissau but the potential application of IPM against important pests of food crops is probably higher in Guinea-Bissau than in many other West African countries. This is because the agricultural practices are unsophisticated with little reliance on pesticides.

In 1984 in Guinea-Bissau biological control was used to control the cassava mealybug *Phenococcus manihoti* with the introduction of the parasitoid *Epidinocarsis lopezi*.

A survey over the cassava growing area revealed the extent of the problem, and IITA accepted Guinea-Bissau as one of the countries affected by this pest. For the first time classical biological control was used to control the pest. Before releasing the natural enemy, farmers were told how biological control differed from insecticide control. After releasing *E. lopezi*, its distribution was monitored and records have shown that 70% of the pest has been controlled by the parasitoid.

This programme was supported by USAID and IITA but currently there is no financial support for the work.

A plan for future work has been drawn up. A survey will be undertaken to monitor the cassava green mite *Mononychellus tanajoa*, which has become as important a pest as cassava mealybug in the country. It is hoped to continue monitoring *E. lopezi* activity and initiate a biological control programme on cassava green mite and mango mealybug *Rastrococcus invadens*. Collaborative work with the DFPV in Niamey on a locust and grasshopper pathogens survey is also underway. At the present time IPM implementation cannot be met by the Government of Guinea-Bissau because of limited resources.

Integrated pest management in Niger

M. MOUNKAILA
Ministry of Agriculture and Stockraising, B.P. 429 Niamey, Niger

INTRODUCTION

IPM in Niger dates from 1972 when, following an evaluation of increasingly expensive chemical control of the white-scale insect *Parlatoria blanchardi*, a pest of date palms, the Biological Control Centre at Agadès was set up. The government gave agricultural departments responsibility for developing cheaper methods for plant protection. With technical assistance provided by IRFA-CIRAD, the aim was to control the white-scale insect by making use of its natural enemies.

The drought years 1973 and 1974 in the Sahel increased food shortages in the region, resulting in the setting up of CILSS. A CILSS plant protection programme was initiated with the technical support of FAO. IPM formed part of the approach adopted by the programme, with the primary objective of establishing research structures for integrated control in member states. In Niger, this facilitated the extension of biological control to pests of crops, including millet, sorghum, maize, cowpea, groundnut and market garden crops.

IMPLEMENTATION PROCEDURES

In the case of the white-scale insect, the initial aim was to protect its local, natural enemies by preventing the use of pesticides in the palm groves. The use of these natural enemies followed an inventory. Because the results were not satisfactory, the new, auxiliary insects were introduced in 1973. These consisted of 4000 coccinellids (*Chilocorus bipustulatus inranansis*) originating from the IRFA laboratory in Mauritania.

After these had acclimatized to their new release area, their predatory effectiveness was compared with that of local predators. Research included a number of scientific investigations based on observations, particularly several releases in the palm belt, and in the laboratory, where biological properties were investigated, and the mass rearing of insects was carried out.

The health of a number of individual insects was monitored over several months. Thus, for each species of predator insect studied, the following observations were made:
- investigation of climatic conditions in Niger
- investigation of the entomological ecosystem in the palm groves and natural enemies
- the biological cycle; lifetime and relationship to the pest's (host insect's) life cycle
- ecological requirements
- proliferation and fecundity in relation to the host insect
- specificity and food preference at different stages of the pest's life cycle
- parasite and hyperparasite research and monitoring.

The IPM programme dealt with crop pests of regional importance – for example, millet, sorghum, rice, maize and market garden crops. In Niger, the programme concentrated on millet and sorghum. A pilot programme for millet crop protection was initiated. This involved farmers, from sowing to harvesting, using the following methods:
- treating seeds with thioral
- agricultural timing and methods
- the choice of improved varieties depending on area of cultivation
- monitoring enemies in pure cultures
- selective use of pesticides
- uprooting *Striga hermonthica* plants
- intercropped millet/bean cultivation.

All these methods were developed from research work on the cultivation of millet carried out at the National Institute for Agricultural Research in Niger (INRAN).

The main crop pests investigated were:
- the white-scale insect (*P. blanchardi*) of date palms
- the millet head-miner (*Heliocheilus albipunctella*)
- the sorghum midge (*Contarinia sorghicola*)
- insects affecting the main food crops, millet, sorghum, maize, cowpea and market garden crops
- millet downy mildew (*Sclerospora graminicola*), *Striga hermonthica*, and the sorghum smut, *Sphacelotheca ehrembergii*, and *Striga gesnerioides* of cowpea.

74

The programme involved developing new techniques for crop protection based on protecting the environment, and avoiding the disadvantages of chemical control, resistance to pesticides and other toxic effects.

These problems were identified by research workers in Niger following research into the frequency, significance and geographical distribution of the pests. The solutions adapted were based on the establishment of crop pest monitoring networks, and the use of improved or resistant varieties produced by research at regional level. This implied a careful choice of pesticides, with the prohibition of some products (DDT, HCH in particular) and the introduction of insecticide treatment in areas (for example, northern palm belt) where biological control was already established.

IPM was selected on the basis of experience in developed countries, particularly the US, where the conventional chemical remedy (DDT), had proved ineffective and gave rise to ecological imbalances. The fragile nature of the Sahel agro-ecosystem, caused by years of drought, has brought about an awareness of the need to protect the environment, and the danger of using pesticides alone.

DISSEMINATION

Farmers were involved in the pilot tests, in monitoring crop pests and in plant health treatments through the work of village brigades. The dissemination service of the Plant Protection Department (Departement Protection Vegetaux, DPV) was responsible for directing the project, and was, therefore, closely involved in the identification of the problem and the preparation of projects. Management and implementation of the project was supervised by the DPV. Insecticide treatments were also controlled by the DPV which monitored the products and dosages used.

Conventional methods of control were disseminated by extension workers. The DPV is continuing the research programme and disseminating the results.

CONSTRAINTS

A number of factors acted as constraints on the successful implementation of IPM:

(a) difficulty in evaluating the extent of losses resulting from plant pests and, therefore, being able to determine their importance – this was a key factor in the difficulty of obtaining funds;

(b) difficulties in evaluating the impact of plant health control measures on agricultural production;

(c) the popularity of chemicals in an environment where farmers, developers and all those concerned are determined to safeguard food production at any cost;

(d) the fact that, by its very nature, research does not yield quick results;

(e) climatic unreliability which made it difficult to carry out pilot projects.

In spite of those constraints, however, there was an awareness of the need to protect the environment.

RESULTS
Biological control of the white-scale insect (*P. blanchardi*)

Here, the results have been encouraging. The insect predators introduced became well-acclimatized and were observed from 1973 onwards. Wherever predators have become established throughout the palm belt, the population of scale insects has fallen considerably and continues to fall. However, no biological control operation can entirely eradicate pests. The Biological Control Centre has also been active in carrying out research into predators of the millet head-miner and other market garden crop pests. This programme has already made an encouraging start and a large number of predators and parasites have been identified, while others are in the process of being identified.

Chemical control

Chemical control involves the application of micronized sulphur mixed with a fine dust (collected in Irhazer) to treat an infestation by a species of mite, *Olygonicus* sp., which affects 85% of the date harvest. The results achieved have been encouraging and satisfactory.

IPM against millet and sorghum pests

The results obtained from biological control are encouraging and will be of great importance in IPM. Parasitic Hymenoptera and predatory Coleoptera (Coccinellidae) (predators of aphids and mealybugs) are being raised in INRAN's laboratory at Agadès under satisfactory conditions. Given the importance of the results obtained, and the potential economic role which these parasites and predators could play in restricting crop pests, it is essential that these useful insects be preserved.

The entomological research work undertaken by INRAN needs to continue in order to provide Niger's agricultural services and its auxiliaries with a valuable tool for effective control against pests, and for maintaining the biological equilibrium of the environment.

Integrated pest management in Nigeria

S.M. MISARI, A.M. EMECHEBE, T.N.C. ECHENDU and S.T.O. LAGOKE
Institute for Agricultural Research, National Root Crops Research Institute, Nigeria

INTRODUCTION

IPM can be described as the reduction of pest problems through the combination of several control methods in a complete programme. Pest-crop interactions must be understood, and ecological as well as socio-economic factors are important to maintain pest pressure below damaging levels, but not eradicate them.

IPM is not a new concept in Nigeria. The farmer in Nigeria has traditionally practised cultural methods of control, such as varying the planting date of a crop, and used good farming practices, such as selecting seed from plants which have survived pest attack. Experience has shown these farmers that mixed cropping is more successful than monoculture or continuously cropping the same piece of land, and that slash and burn with long periods of fallow is an effective way to alleviate pest problems. This shows that farmers have consciously or unconsciously incorporated IPM practices into their production systems as they usually combine two or more of the above practices in their crop production processes.

It has been estimated that only 48% of the land available for cultivation in Nigeria is used with 64% of the working population engaged in agriculture. The farming population is almost all small-scale farmers with little capital outlay, dependent on the human labour force. It cannot produce enough food for the population, recently estimated at 88.5 million. Pest problems are increasing and, with arable land a limited renewable resource facing competition from non-agricultural uses, increases in crop yields through effective IPM programmes have been identified as the answer to the food and nutrition problems in Nigeria. We attempt here to discuss the institutional infrastructure, policy implementation and constraints working against IPM in crop production research and development efforts in Nigeria.

RESEARCH AND DEVELOPMENT OF IPM

Institutional networks

IPM research and development programmes are carried out by three different but complementary sectors in Nigeria. The universities and polytechnics carry out basic and applied research and the research institutes are concerned with basic, applied and adaptive research. Both universities and research institutes collaborate with IARCs, such as IITA, ICRISAT, CIMMYT, WARDA etc. The third sector is the State Agricultural Development Projects, and the State and Federal Departments of Pest Control Services (FDPCS).

The FDPCS and the State Plant Protection Services are expected to work together with research institutes to develop sustainable IPM packages, but they have been more involved in chemical pest control, especially of locusts, grasshoppers and quelea birds. The FDPCS is non-governmental and private organizations, such as chemical companies, are involved indirectly with the government research institutions who test the products with potential for IPM programmes.

In Nigeria there are currently 21 federal universities with faculties of agriculture and nine state universities. In addition there are two agricultural universities and 18 agricultural research institutes, each with a clear and specific research mandate. All these institutions are involved in biological and pest management research. Research in each of these institutes is based on crop commodity programmes and farming systems. Each project is carried out by a multidisciplinary research team, and approval and priority rating rests with the respective multidisciplinary research committees.

Institutional and government involvement and commitment

Of the six areas of crop commodity research which comprise varietal improvement, cultural practices and management, crop protection, post-harvest technology, environmental effects, and the socio-economics of production, crop protection takes as much of 20–25% of the funds committed to implementation. At least 90% of the funds for research in terms of personal emoluments, infrastructure and supporting services are from the government. Each university and research institute is allocated a proportion of recurrent expenditure depending on the size of the priority programme. The level of funding may be substantial depending on the pest problem and its prominence at a particular period. Little bilateral funding is available for agricultural research and the research. The government is committed to IPM for politically sensitive cash crops such as cocoa, groundnuts and cotton which receive special support.

Food crops such as cassava, maize and cowpeas have also received attention with respect to IPM. Projects for the control of cassava mealybug, mango mealybug, African gall midge, black sigatoka disease, rice blast and water hyacinth have been funded by the federal government.

IPM PROGRAMME

IPM constitutes an integral part of curricula in faculties of agriculture and programmes of reseach institutes. Initial problems are identified through farmers, field reports, extension staff and research workers. Indirectly through monthly technology review meetings of the train and visit system of extension, research workers get to know farmers' problems, visit farms and become familiar with on-farm adaptive research activities.

Numerous problems are being addressed depending on the crop, geo-political or ecological zone and the priority rating and level of funding attached to them. A few problems of national significance are groundnut rosette disease, *Striga* on maize, sorghum and millet, sorghum downy mildew on maize, water hyacinth etc.

Examples of such collaborations include Institute of Agricultural Research (IAR), Samaru,/IITA/ FAO work on *Striga*, SAFGRAD networks, IDRC; others include the IAR-Samaru/ICRISAT/ WCAMRN research on millet downy mildew and National Root Crops Research Institute, Umundike/IITA work on cassava mealybug and green spider mite etc. Although such relation-ships are essentially collaborative, there are others that comprise financial and logistic support including training and information exchange. We appreciate such support from FAO, SAFGRAD, IDRC, ODA and other such organizations.

Development of IPM research strategies

Once a problem is identified and approved as a research project, the research strategy is first tested on-station for 2–3 years, which involves adequate exposure to pests and diseases. After the second or the third year of on-station testing, it is ready for testing at farm centres. This takes at least two more years depending on the incidence of pest and diseases. Concurrently, if the results from the multilocational trials appear repeatable, the technology goes into on-farm adaptive research or farmers' fields. Thus a combination of appropriate IPM components are put into packages for testing on farmers' fields with farmer participation and researcher-managed trials initiated; later the activity is converted to researcher participation and farmer-managed.

Once this technology is successful at the on-farm level, it is then scrutinized by a research review committee before final ratification and release by the research professional and academic or governing board as a recommendation for general consumption. Then the agricultural extension and research liaison services can disseminate the information through the publication of extension bulletins, field extension training, radio broadcasts and other media.

IPM training

Middle-level manpower training in general agriculture is conducted in colleges of agriculture which run a programme specifically on crop protection up to Higher National Diploma level. In-service training on specific crop protection topics are also organized for field extension staff by the agriculture extension and research liaison services at a local (or state) and national level. Training up to higher degree (M.Sc.) level in crop protection is also offered by various universities. Short and medium-term refresher courses or higher degree training in IPM-related courses are sometimes available overseas but this requires external sponsorship.

IPM results

Progress has been made in developing and recommending IPM packages for the control of some major pest problems in the country. A combination of tolerant varieties, fertilizer and herbicide have been used to control *Striga hermonthica* in sorghum in the Sudan zone and in maize in the Northern Guinea Savanna at farmer level. Similarly, early sowing, use of resistant material and effective seed treatments using metalaxyl gave effective control of downy mildew in maize. Cassava mealybug is being controlled through a combination of resistant variety, fertilizer and the biocontrol agent, *Epidinocarsis lopezi*. In groundnut, the devastating groundnut rosette virus disease is being controlled by the combination of early planting, close spacing, use of resistant varieties and appropriate systemic insecticides. Seed dressing, closed season and appropriate improved variety have been widely adopted for the control of bacterial blight of cotton by most farmers in the country.

Constraints

The major constraints are:

- financial: government funding for IPM programmes is inadequate

- government policy: erratic policy changes towards science and technology does not allow systematic, sustainable research

- manpower: inconsistent government policy has resulted in a lack of well-trained manpower

- technical constraints: scarcity and high cost of inputs such as seeds and agrochemicals which are often poor quality when available; most research has been directed at monocropping systems, while farmer practices are predominantly intercropping, so there is an incompatibility of proven technologies for their production systems

- infrastructure: there is a lack of effective plant quarantine and appropriate facilities for rearing insects, particularly natural enemies for biological control components of IPM

- information resources: poor national documentation services

- farmers attitudes: farmers prefer their traditional practices and are resistant to change.

Further research needs are:

(a) identification of the insect pest problems and their natural enemies in mixed cropping systems under varied environmental conditions;

(b) a study of the ecological factors which regulate the distribution and abundance of pests in mixed cropping systems;

(c) establishment of reliable economic injury levels;

(d) the development of ecologically sound crop protection methods for particular mixed cropping situations, e.g. search for effective and selective chemicals and their proper integration with other control methods;

(e) the development of effective methods of transfer of IPM technology to small-scale farmers;

(f) collaboration with plant breeders and other disciplines to find pest resistant/tolerant varieties and option crop management practices to determine the usefulness of these in reducing crop losses;

(g) All these can only be done if sustainable funding and inter-institutional and international collaboration are available.

Integrated pest management in Senegal

M. WADE

Institut Sénégalais de Recherche Agricole, CNRA, Bambey, Diourbel, Senegal

INTRODUCTION

The development of countries in the Sahel is essentially based on the promotion of rural life, and agriculture remains the driving force in this process. The main objective of governments, including the Senegalese Government, is to increase food production and become self sufficient.

As Senegalese agriculture is primarily subsistence agriculture, some methods or strategies for the control of crop pests can be ruled out. Experience of popularizing methods of crop protection in Senegal reveals four factors which may delay the establishment of IPM in subsistence agriculture: technical, economic, sociological and institutional.

TECHNICAL ISSUES

Technical issues as they relate to IPM involve four principles: use of pesticides, agricultural practices, resistant varieties and biological control.

Use of pesticides

Most subsistence farmers have received little schooling, and many are illiterate. Any product which is hazardous to use must be avoided, because instructions on the labels of pesticides packaging will not be followed. This restricts the choice of usable pesticides. The application of pesticides is often difficult (height of plants etc), particularly if farmers are using manual equipment.

Agricultural practices

The system of land tenure in Senegal, particularly that of the most populated areas, undoubtedly acts as a brake on the development of IPM. For example, some useful practices – a certain type of rotation, eliminating some plants so as to reduce their seed production, ploughing at the end of the cycle (an effective control against *Heliocheilus chrysalises*) – are ruled out. A technique which is known to be effective in reducing infestations of *Striga hermonthica* – manuring by cattle – will only be carried out by farmers on a plot which they can be sure of working for several years.

Experience of experimental units, concerned with advising rather than instructing farmers, has led to the setting up of some permanently stable, individual holdings.

Because the rainy season in Senegal is short, it is difficult to find a sowing date on which a sensitive stage in the crop's development does not coincide with the period when a parasite is present or abundant. Neither is it possible to bring forward the date of sowing because most of the seeds, for example, millet, are sown dry before the start of the rains, or once these have begun.

Associate crops, although beneficial in reducing populations of crop enemies (*Striga* and insects) are almost non-existent, because they often cause difficulties if cultivation has to be intensified and mechanized.

Despite the decimation of cattle stocks following years of drought, farmers do not favour this practice for two reasons: the fertilizing effect, which is of little value if there is not enough rain during the year; and the transhumance of cattle after harvest through lack of forage.

Resistant varieties

Research into resistant varieties has yielded useful results. However, some varieties have not been accepted by the farmers, for example: variety b-301, which is resistant to *Striga gesnerioides*, and GAM millet, because of its low height, the straw is valued and can be used for other purposes.

Biological control

As far as biological control is concerned there has been some progress, the most encouraging being

80

the use of *Bracon hebetor* to control *H. albipunctella*. In some rainy areas (500 mm or more), *Smicronyse* has proved to be very effective against *Striga hermonthica*.

In recent years interruptions in rainfall of two to three weeks have been recorded at different periods during the growth of the crop in many areas. These conditions are often unfavourable for the use of some natural enemies for pest control.

ECONOMIC CONSTRAINTS

Many Senegalese farmers earn very low incomes and this limits their capacity for investment. This means that it is not sufficient to demonstrate that a control strategy is viable. Its applicability also has to be evaluated. The uncertainty over future rainfall may also give rise to uncertainty about the economic viability of, for example, introduced species.

SOCIOLOGICAL CONSTRAINTS

The essential difficulty in persuading farmers to accept IPM is that their perception of plant health problems differs from that of research workers. It is not easy to reconcile these differences. There is the added constraint that because farms are, generally, not highly technological, some methods of pest control are not readily acceptable. To date, most pest control in subsistence agriculture such as is found in Senegal has consisted mainly of chemical application. With pesticides, the effects are easily seen and give the farmers confidence. This means that it will be difficult to persuade farmers to adopt other control techniques where the results are harder to perceive.

INSTITUTIONAL CONSTRAINTS

Despite the existence of legislation on plant health, vulnerable regions in Senegal are not yet protected from new pests because the legislation is not vigorously enforced. The plant protection services are mainly concerned with insect and avian pests. Furthermore, if integrated control requires introduced species, two problems may arise: the availability of species at the required time and place; access to credit to enable the farmers to buy the species.

The National Programme for Agricultural Popularization is attempting to solve these problems. Although still largely undeveloped, this is the appropriate body to popularize simple technology among the farmers. Its activities, however, will need to be reinforced by those of the agricultural bank, CNCA. The NGOs also have a role to play in eliminating some of the constraints discussed above.

Discussion

The first issue raised for discussion by a representative from Ghana was the need to strengthen IPM infrastructures at the national level. He commented that local governments and institutions must take greater responsibility for promoting long-term IPM sustainability, especially in providing adequate training for national scientists rather than bringing in foreign specialists. The speaker from Niger added that in his country there was still inadequate staff in IPM and that training was a serious problem. No IPM staff in Niger are eligible for further training, due to a rule that only government staff with five years work experience can receive training.

In response to the country report from Nigeria, K. Cardwell commented that the IPM for downy mildew in maize described by Dr Misari had, in fact, not succeeded in being adopted by farmers due to the costs involved and the marketing difficulties arising from early planting etc. She added that the availability of resistant varieties was also limited due to a lack of national viable seed production, and suggested the promotion of private seed companies or the provision of agricultural credit for seed production as two possible solutions to this problem. Dr Misari agreed with Dr Cardwell's remarks, adding that the lack of seed production in Nigeria was linked to a lack of political will and that the 1991 downy mildew epidemic was actually the result of a number of interrelated factors.

Finally, A. Dreeves suggested that issues for further discussion should include the purposes and benefits of working with farmer groups, as had been mentioned in several presentations; the most effective strategies for reaching farmers at the village level; and the best mechanisms for changing attitudes towards pesticide use and reducing extensive spraying. The representative from Senegal agreed, adding that the most important challenge in IPM is for the researcher/practitioner to put themselves in the farmers shoes and attempt to see things from their point of view.

Integrated pest management in Zaire

H.D. NSIAMA SHE
Ministry of Agriculture, Kinshasa, Zaire

INTRODUCTION

The main food crops in Zaire include cereals (maize, rice, sorghum), root and tuber-bearing plants (cassava, sweet potato, yam, potato), legumes (soyabean, green bean, cowpea, groundnut, chickpea), bananas and plantains. Almost 35% of the land under cultivation is devoted to cassava. Cassava provides more than 70% of the daily calorific needs of more than 60% of the population. The leaves are used as vegetables and represent approximately 60% of Zaire's annual vegetable production.

Cassava is the basis of the cropping system and is associated with a wide variety of crops, (for example, beans, sweet potato, maize, sorghum, pepper, tomato, sorrel, goundnut, squash etc.). Investigation and understanding of the productivity of cassava-producing eco-systems (cassava and other crops) will contribute to an understanding, and the management of food crop agricultural systems, including issues of plant protection. Furthermore, because cassava occupies almost 1% of the country's land area, this gives it a major ecological role, which may be affected by pests.

The cassava mealybug, *Phenococcus manihoti* (Homoptera, Pseudococcidae) and the cassava green mite, *Mononychellus tanajoa* (Acarina, Tetranychidae), are two pests which were accidentally introduced into Africa from Latin America in the early 1970s. The two pests were first recorded in Zaire around 1973 (Hahn and Williams, 1973). Both, and the cassava mealybug in particular, have caused and continue to cause damage and considerable economic losses (Hennessy *et al.*, 1990). This has led to socio-economic and political consequences. Being exotics, conventional biological control was prescribed for the two pests (Anon, 1978; Greathead, 1978).

INSTITUTIONS INVOLVED IN THE DEVELOPMENT OF IPM

Through its national programme of agricultural research into cassava (PRONAM), the government has overall responsibility for the development of IPM. While the NGOs are not, in general, familiar with IPM, some NGO officials have indicated that they might become more actively involved if programmes for training, organization, co-operation and research existed. The universities do not offer specific training in IPM. Those involved in the development of IPM include the few entomologists in the national programme. Research topics are developed making use of the concepts and philosophy of IPM. Pathologists, weed experts and other research workers agree that their work should, in future, be better integrated.

One result of the lack of co-ordination on education is that few research workers receive special training in IPM. However, their training in agriculture, biology, ecology and other disciplines, provides most researchers with an adequate knowledge to use the concepts and philosophy of IPM in team work. Most are of foreign-trained M.Sc. or Ph.D. level.

No local funds are purely devoted to IPM research. The government and USAID provide funds for the general management of PRONAM. Almost all the funds for practical biological control come from the Biological Control Programme of IITA.

Biological control work is carried out as part of the entomology section in the same way as work into resistance of the host plant. A restructuring of SENARAV may see IPM emerge as a department or a division. Recommendations in the form of leaflets are generally produced for the integrated utilization of agricultural research data obtained in different fields of cassava production. Up to now, protection measures have been indirectly put into practice in this form.

ESTABLISHMENT AND CONTACTS WITH OTHER INSTITUTIONS

The idea of IPM comes from the recommedations of a 1977 seminar on the cassava mealybug held in Mvuazi, Zaire (Anon., 1978), to which both national and foreign specialists contributed. The cassava mealybug was identified as an ideal candidate for conventional biological control (Greathead, 1978), but other approaches were also suggested (Anon., 1978).

The Mvuazi seminar recommended conventional biological control and co-operation between national institutions (the Plant Protection and Quarantine Departments in Zaire and PRONAM) and international institutions (IITA, IIBC, IDRC and University of California). Later, USAID was to finance activities at a national level. Other international bodies such as IITA provided scientific and technical personnel or funds for the investigation and rearing of natural enemies, release and follow-up operations and various forms of training, as well as the costs of attendance at seminars and conferences.

LOCAL STRATEGIC INVOLVEMENT

In addition to the work carried out by PRONAM, an official policy on biological control as a basis for IPM has been in place since 1990. The Plant Protection and Quarantine Service are keen to see plant protection for other crops evolving towards this environmentalist approach, as in the case of cassava. IPM is well reflected in the research being carried out in two other national programmes on maize and legumes, both of which have recently been orientated towards IPM, based on the PRONAM model. The determining factor here has been the experience and formidable success achieved with biological control against the cassava mealybug. This was the result of extensive co-operation between different international institutions (Herren, 1989) and the acquisition of skills by PRONAM.

Methodological approach

Co-operation research work which followed the Mvuazi seminar identified South America as the origin of the cassava mealybug (Matile-Ferrero, 1977). This led to research aimed at investigating the mealybug's natural enemies, importing them and releasing them in Zaire. In addition to this conventional biological control operation, other control tactics, resistance of the host plant, agricultural practices, sanitation etc., were included in the IPM programme being developed (PRONAM, 1978; Nsiama She *et al.*, 1990). Following this work, PRONAM is now well placed to provide a number of formulae for integrated biological control of the cassava mealybug and the cassava green mite (Nsiama She *et al.*, 1990).

However, progress was not straightforward. Because the cassava mealybug seemed set to spread outside Zaire, the international community took responsibility for financing the research. However, this could not begin immediately because the scientific expert at PRONAM lacked expertise in biological control. His training, and that of other personnel, delayed implementation. Lack of infrastructure, for example, laboratories, also hampered operations. Poor systems of transport and communication also slowed down developments. For example, in 1978 all imports of natural enemies to control the cassava green mite in the Ruzizi Valley were lost following delays and cancellations in air transport between Kinshasa and Bukavu. Another consignment was finally released at a level of between 1% and 5% because of poor pre-release investigation work and lack of communication between personnel involved in the operation.

Involvement of farmers

Up to the beginning of conventional biological control against the cassava mealybug, farmers had been little involved in the search for solutions. From 1983, however, PRONAM began to adopt research programmes which looked at farming systems and agricultural activity. This increased the involvement of farmers. The opinions of farmers were collected through questionnaires, which were converted into research topics. From 1988, farmers were increasingly involved through pilot tests. In 1990 PRONAM restructured its research programme with the aim of further involving farmers. The aim was to investigate the complex of plant health, agricultural, socio-economic and environmental parameters on the ideal type of cassava for farmers. The hope was that this approach would contribute to identifying means of shortening the cassava selection cycle and might also produce an improved cultivar which would have a high probability of being accepted by farmers.

As far as farmers themselves are concerned, many remain ill-informed as to the nature of cassava infestation. Farmers also lack organization and are, generally, poorly managing crop pests.

Zaire's Plant Protection and Quarantine Departments play only a symbolic role in the development and implementation of IPM. The Departments authorize the importation of beneficial insects, at the request of PRONAM. The Service National de Vulgarisation Agricole (SNV) is still very new. Fruitful co-operation in the form of joint work on cassava in pilot studies, already exists between SNV and PRONAM/SENARAV. Both the Plant Protection and Quarantine Departments and SNV need to be strengthened through training, fieldwork and continuing dialogue with PRONAM and SENARAV. For example, the Plant Protection and Quarantine Departments have accepted donations of insecticides, although they have neither the personnel nor the infrastructure to test the ecological reliability of toxic products. PRONAM and SENARAV, which are relatively better equipped, could help the Plant Protection and Quarantine Departments to make better environmental choices on foreign aid, within the framework of the National Plant Protection Programme.

RESULTS AND PROSPECTS

The programme of biological control against cassava mealybug has been and remains successful. Cassava is again being cultivated where previously it had completely disappeared, as in southern Shaba. The programme is continuing because new outbreaks of cassava mealybug are again being reported. New releases of the most promising natural enemies are scheduled. The selection of resistant varieties is progressing and there is promise of multiple resistance to diseases and insects (Nsiama She, 1990).

The data obtained in Zaire and elsewhere (Neuenschwander *et al.*, 1989) on the conventional biological control of cassava pests may act as a model for optimization techniques for the management of cassava eco-systems in Zaire, and as a starting point for the application of integrated pest control to other crops. An IITA-Zaire project financed by GTZ began with this object in 1991. Teams of research workers from different ecological, agricultural and social disciplines will study food crop pests in order to develop better forms of management.

Through a vigorous training programme under the aegis of USAID, IITA and FAO, several national scientists have been trained in many fundamental disciplines for the design and development of IPM systems.

REFERENCES

ANON. (1978) Recommendations. pp. 80–85. In: *Proceedings of an International Workshop on the Cassava Mealybug* (Phenacoccus manihoti *Mat-Ferr* (Homoptera: Pseudococcidae)), *INERA, Mvuazi, Zaire, June 1977.* NWANZE, K.F. and LEUSCHENER, K. (eds).

GREATHEAD, D.J. (1978) Biological control of mealybugs (Homoptera: Pseudococcidae) with special reference to the cassava mealybug (*Phenacoccus manihoti* Mat-Ferr). pp. 70–80. In: *Proceedings of an International Workshop on the Cassava Mealybug* (Phenacoccus manihoti *Mat-Ferr* (Homoptera: Pseudococcidae)), *INERA, Mvwazi, Zaire, June 1977.* NWANZE K.F. and LEUSCHNER, K. (eds).

HAHN, S.K. and WILLIAMS, R.J. (1973) *Investigations on Cassava in the Republic of Zaire.* Report to the Commissaire d'Etat à l'Agriculture. March 1973. (unpublished)

HENNESSEY, R.D., NEUENSCHWANDER, P. and MUAKA T. (1990) Spread and current of the cassava mealybug (*Phenacoccus manihoti* Mat-Ferr (Homoptera: Pseudococcidae) in Zaire Tropical. *Tropical Pest Management*, **36** (2): 103–107.

HERREN, H.R. (1989) Le programme de lutte biologique de l'IITA: du concept à réalité. pp. 20–33. In: *La Lutte Biologique: Une Solution Durable aux Problèmes Posés par le Déprédateurs des Cultures en Afrique. Actes de la Conférence et de l'Atelier d'Inauguration du Centre de Lutte Biologique de l'IITA pour l'Afrique, Cotonou, Bénin, Décembre 1988.* YANINEK and HERREN, H.R. (eds).

MATILE-FERRERO, D. (1977) Une cochenille nouvelle nuisible au manioc en Afrique Equatoriale *Phenacoccus manihoti* n. sp. (Homoptère: Pseudococcidae). *Annales de la Societé Entomologique de France*, **13**: 145–151.

NSIAMA SHE, H.D., LUTALADIO, N.B. and MAHAUNGU, N.M. (1990) Integrated management of cassava pests in Zaire. Status and prospects. In: *Proceedings of the Global Status and Prospects for Integrated Pest Management of Root and Tuber Crops in the Tropics, Ibadan, October 1987.* HAHN, S.K. and CAVENESS, F.E. (eds).

PRONAM (1978) Programme National Manioc. *Rapport Annuel.* Kinshasa: Ministère de l'Agriculture.

Integrated pest management in Burkina Faso

T. DOULAYE

DPVC Laboratoire de Kamboinse, B.P. 5362 Ouagadougou 01, Burkina Faso

INTRODUCTION

An overall approach to IPM was initiated in Burkina Faso following the establishment of the integrated control project financed by USAID through CILSS. The project began in 1983 and ended just after its first phase in 1987 following a change in US policy.

The project financed the building of three research laboratories, as well as providing specialist training in the US and France in entomology, weed research and plant disease. The laboratories are, at present, being used by local researchers for programmes which are financially and technically supported by CIDA. Research is managed jointly by the Canadian Agriculture Research Station at Saint-Jean-sur-Richelieu (Canada) and the research stations at Bobo-Dioulasso and Kamboinse (Burkina Faso).

Local involvement is reflected in the many high-level personnel who were made available by the Department of Plant Protection and Packaging (DPVC) for both research and practical programmes.

HOW PROGRAMMES ARE PLANNED AND EXECUTED

Crop yields in Burkina Faso are reduced mainly because of damage caused by pests such as grasshoppers, locusts, stem borers, spike suckers, millet head-miners, sorghum midge and the parasitic weed, *Striga*. The DPVC is decentralized into 11 plant health bases. This means that each base is in constant contact with farmers, helping them to combat pests.

Generally, control treatment begins with chemicals which have been tested for their effectiveness. The next stage involves research projects with a view to introducing IPM. Biological control is favoured because it is cheap and viable in the long term for subsistence agriculture. It also poses no danger to the environment, an important factor in a fragile and threatened environment.

The farmers are involved at all stages of this process. All research projects are formulated on the basis of the needs of the farmers, who are also involved in pilot studies. The aim is to enable farmers themselves to apply the technology resulting from research. The DPVC's training and popularization department is involved in training at all levels. The aim is to train 11 000 farmers each year, enabling them to monitor and control the pests that are affecting their crops. The DPVC works in close co-operation with the National Popularization Department in Ouagadougou.

Methodological approach

IPM in Burkina Faso is based on two approaches:

(a) biological control of the millet head-miner *Heliocheilus albipunctella* (Lepidoptera, Noctuidae);

(b) biological control of *Striga hermonthica* and *S. gesnerioides*. Whereas the former has been subject to field tests, the second programme is just beginning.

Biological control of the millet head-miner

Heliocheilus albipunctella is parasitized by *Bracon hebetor* (Hymenoptera, Braconidae) and parasite levels vary between 2% and 14%. A method of raising *Bracon* was developed in Senegal and releases were made by V.S. Bhatnagar of FAO, as part of the biological control programme of the CILSS IPM project. FAO technicians were made available to other CILSS integrated control project research workers, in an attempt to determine the optimum number of alternative hosts, *Ephestia* sp., for *Bracon*.

The work was carried out at farmer level in Ouahigouya. The aim was to collect natural populations of larval parasites of *Heliocheilus*, including *Bracon hebetor*, and to determine their potential effectiveness and the possibilities for their use in biological control.

Larvae of *Ephestia* sp., raised for this purpose in the laboratory since 1987, were used to attract females of the parasite looking for a laying site. Pieces of hollow bamboo (15 × 5 cm) with lateral holes and provided with fine muslin cloth sleeves, containing the *Ephestia* larvae and millet flour, were suspended in a shed with a straw roof.

Four treatments with seven repetitions were used (T0 = control, T1 = 15 larvae, T2 = 30 larvae, T3 = 45 larvae). The larvae were exposed to attack by *Bracon* for 21 days, after which the percentage parasite infestation was estimated. Parasitized larvae were then kept in the laboratory until the parasites emerged, and these were used to raise more, again on *Ephestia* larvae.

Biological control of Striga

From 1987 research work was carried out at the Kamboinsé laboratory to determine the impact of insects on *Striga* and to identify potential agents for biological control. These preliminary studies carried out between 1987 and 1990 were concerned with the potential of *Smicronyx* spp. (Coleoptera, Curculionidae) for the biological control of *Striga*.

Ten unit plots of 10×4.5 m were randomized into five Fisher blocks. Each block consisted of two plots, one of which was treated and the other untreated. Total exclusion of the insects were possible through the application of diazinon every 10 days at a dose of 1.5 l/ha following the appearance of the first growth of *Striga*. All the plots received 100 kg/ha of NPK on sowing, and 50 kg of urea on the 50th day after sowing.

At the end of the cycle all the *Striga* stalks were cut flush with the ground and automatically weighed to determine the weight of fresh material in each plot. After drying for 5 min in a stove at 70 °C they were weighed again to obtain the weight of dry matter.

In order to determine the damage caused by gall-forming insects, *Striga* plants collected from the two central rows in each individual plot were separated into two groups of attacked and unattacked stems. Each group contained the same number of stems.

Insects found on Striga

All the insects present on the *Striga* stems at all stages were collected. Likewise, the parts of the host plant on which the insects were frequently observed were noted. Insects captured were classified by order and family, and are subdivided by numerical importance, as follows:

Coleoptera = 54.42% captures. Three families were found:
 Curculionidae = 72.07% of the total for this order
 Chrysomelidae = 19.82%
 Meloidae = 1.80%

Heteroptera = 24.51% of captures. The following families were identified:
 Pentatomidae = 40% of the total for this order
 Ligaeidae = 35%
 Coreidae = 8%
 Miridae = 6%

Lepidoptera = 7.35% of all those collected. The following families were identified:
 Satyridae = 60% of the total for this order
 Noctuidae = 26.66%
 Geometridae = 13.33%

Hymenoptera = 7.35% of total captures. In this order the insects found mostly belonged to the family Formicidae, representing 93.33% of all those counted.

Orthoptera = 4.41% of captures. The families found were:
 Pygomorphidae = 33%
 Catantopinae = 11.11%

Neuroptera = 1.47% of total captures.

Results

The role of many of the useful species may vary greatly depending on the year, season and crop, as well as on factors such as climatic conditions, or whether or not the periods when the pest and the predator appear occur simultaneously, and the likely presence of hyperparasites.

The programme for increasing *Bracon hebetor* had to be suspended when some of the research workers departed for training. Because this programme has a contribution to make in the control of *Heliocheilus*, the way in which *Bracon hebetor* is collected needs to be improved.

Many insects are present on *Striga*, either to feed at the expense of the plant, or to find other insects, or for other, as yet unknown, reasons. The visible damage is mainly due to coleopteran and lepidopteran larvae, and affects the leaves, flowers and capsules.

The fresh material and the dry matter in *Striga* stems investigated in 1988 and 1989 showed a significant difference depending on whether the plots were treated or untreated. This appears to underpin the view that such insects could be envisaged as biological agents for the control of this weed.

CONCLUSION

Because IPM is in its early stages in Burkina Faso, the main constraint is a shortage of research workers. Biological control does not yield spectacular results and involves long-term work. This means that both practical help and financial support are essential to the success of the approach.

Integrated pest management in Sierra Leone

B.D. JAMES and C.B. SESAY

Ministry of Agriculture, Forestry and Fisheries, Freetown, Sierra Leone

INTRODUCTION

Approximately 70% of Sierra Leone's 4 million population is engaged in agriculture (Anon., 1971). Individual farmers practise a high degree of diversification in crop production, but largely at subsistence levels. The major food crops, rice, cassava, sweet potato, maize, legumes and vegetables, are generally cultivated in mixed cropping systems, whilst cash/plantation crops, e.g. cocoa, coffee and oil palm, tend to occur as pure stands. Agricultural productivity is relatively low in the country, and in all cropping systems crop damage caused by insect pests, disease and weeds significantly contribute to this low output.

The Crop Protection Service (CPS) of the Ministry of Agriculture, Forestry and Fisheries is the main body that handles crop protection matters. Originally it comprised two separate units, the Pest Control Branch, established in 1966, and the Phytosanitary Unit set up in 1974. In 1983 these bodies merged into the CPS. Their mandate is to provide:

- technical advice to farmers on all aspects of crop protection

- a pest control service, or assistance to farmers to undertake pest control operations on their farms

- phytosanitary inspection services and administer plant quarantine regulations at the country's border posts.

Until the mid-1980s the CPS was poorly equipped and had a manpower force that was disproportionately high relative to its functions and operations. The service made up 22% of the entire Ministry's agricultural extension staff, and consisted of 305 field staff and 52 plant quarantine personnel (FAO, 1991a). The head of the CPS was the only graduate in the service at that time, and his duties were mainly administrative. Under his supervision were 17 Agricultural Instructors who had received intermediate level training in general agriculture locally at the Certificate Training Centre (CTC), Njala University College. Almost invariably the Agricultural Instructors relied largely on the knowledge and skills acquired at the CTC in the execution of their duties, further training beyond CTC (either formal or in-service) being rare for the majority of them. By 1986, six of them had received short-term overseas or regional training in crop protection. The need for a stronger CTC crop protection syllabus to reflect the importance of the centre's curriculum in IPM understanding, development and implementation has only recently (1990) been met.

The majority of CPS field staff are agrotechnicians/pesticide operators who lack training in crop protection beyond an initial six-month agronomy course given on appointment. They operate under the direct supervision of the Agricultural Instructors.

Clearly the CPS lacked the capability to fulfil its mandate effectively, and up to the mid-1980s chemical control remained the first option for most crop protection problems. The deficiency in trained manpower resulted in the absence of much needed information for the development of IPM in the country. Crop protection units set up by NGO-supported projects, e.g. various integrated agricultural development projects, and by parastatals, e.g. the Sierra Leone Produce Marketing Board, also relied on the CPS for staff recruitment, compounding the problem. These private crop protection units have, however, faded, along with the phasing out of the projects or with the waning fortunes of the parastatals. The Rice Research Station, Rokpupr, continues to maintain a crop protection department which has over the years conducted basic and applied research geared towards pest management in rice. Increased graduate level staffing and funding are still needed to enable the further development of IPM at the station. Individual researchers in the university also contribute to pest management research in the country, but on an *ad hoc* basis.

The staffing position and training programmes within the CPS had improved by the end of 1990 (FAO, 1991a). The number of graduate staff increased to seven and the number of Agricultural Instructors increased to 21. Fifteen of these had received short-term overseas training. A number of agrotechnicians/operators were dropped and the overall CPS staff reduced to 250. A series of in-service and short-term overseas training was conducted for the graduate staff and Agricultural Instructors on various crop protection topics. These developments were funded and supervised by a UNDP/FAO project which also improved the overall infrastructure of the CPS. A working atmosphere has now also been created with informal linkages between the CPS and the university, as well as between the CPS and the National Agricultural Research Institute (Rice Research Station and Institute of Agricultural Research, Njala). The CPS is now better equipped to undertake its mandate than it was in the past.

IPM INITIATION: CONTACT WITH EXTERNAL INSTITUTIONS

The crop protection search has now moved away from chemical control as the first, and often only option, to a situation of proper problem identification and consideration of alternative control measures. The UNDP/FAO project and the IITA Biological Control Programme have helped to reduce over-reliance on chemical control. The UNDP/FAO project was eventually born out of the country's weakness and inadequacy to deal with the African armyworm (*Spodoptera exempta*) outbreak in 1979 (FAO, 1987). From 1986–91 project staff and their national counterparts helped to improve the functions and operations of the CPS; they collaborated with the Desert Locust Control Organization for Eastern Africa (DLCO-EA) EC Regional Armyworm Project, Kenya and NRI to set up a monitoring and forecasting moth trapping network to guide any future armyworm control. They also liaised with the Asian Vegetable Research and Development Centre (AVRDC) to demonstrate IPM of *Cylas* weevils in sweet potato, and the project greatly supported the IITA-initiated biological control of exotic cassava pests. IITA involvement in promoting IPM in Sierra Leone started in 1985 following the country's request to deal with the cassava mealybug (*Phenococcus manihoti*) outbreak (James, 1987; Neuenschwander, 1985). IITA now has a two-year (1991–93) biological control programme funded by GTZ and run by the CPS in Sierra Leone (Anon., 1991). IPM of the variegated grasshopper (*Zonocerus variegatus*) is currently being demonstrated in the country by the CPS through FAO funding (FAO, 1991b).

Internal funding for research is rare and this has hampered work at the Rice Research Station, Institute of Agricultural Research and the university that could eventually lead to the development of IPM packages.

LOCAL POLICY COMMITMENT

An important step that will eventually support the development and implementation of IPM in the country was the setting up of the National Crop Protection Co-ordination Committee in 1988. The Committee comprises policy-makers from the Ministry and representatives from the agricultural research institutions (Rice Research Station, Institute of Agricultural Research), the university and pesticide marketing companies. It was set up by the CPS with the assistance of the UNDP/FAO project, but has never met since its inaugural meeting. A recently constituted National Biological Control Committee could eventually merge with the National Crop Protection Co-ordination Committee and avoid duplication of effort. Despite the lack of meetings the Committee members individually accept that crop protection field activities have to be organized along the guidelines of IPM.

PROGRAMME FORMULATION AND STRATEGIES

Development agencies have helped to improve the knowledge of field staff, but few IPM programmes have been initiated. Exotic pests such as cassava mealybug (*Phenacoccus manihoti*), cassava green mite (*Mononychellus tanajoa*) and mango mealybug (*Rastrococcus invadens*) are the only ones for which well-defined control programmes and strategies are available (Anon., 1991). For these, the methods and techniques of biological control, e.g. pre-release and post-release monitoring surveys, release techniques, and impact assessment were developed by IITA (Benin). During farmer training sessions the negative aspects of using pesticides in crops where biological control is already in place are discussed. Farmers are also urged to avoid recycling the pest-infested planting material and to report new cases or areas of incidence. The selection of pest resistant varieties as an IPM measure is generally fortuitous, as it is primarily made for the yielding potential of the variety.

Control of *Z. variegatus* by IPM strategies is being demonstrated by an FAO project. The methods involve integration of timely weeding/underbrushing in November/December to discourage/kill early instar nymphs, spot pesticide spraying of the early instar nymphs on weeds in crops and surrounding vegetation, and oviposition site identification (April/May), followed by egg pod destruction by October to reduce population recruitment in the following generation. The large-scale demonstrations will involve several neighbouring farms over a wide area to ensure the pre-requisite community participation for long-term success. The programme will avoid pesticide spraying of cassava crops to prevent resurgence of the cassava mealybug for which biological control is operational. The polyphagous nature of Z. variegatus, the widespread distribution of the damaged areas and the failure of individual farmer's control efforts are key features in the decision to adopt an IPM approach in its control.

A wide variety of natural enemies are associated with insect pests of rice (Azim, 1988). The Crop Protection Division advises farmers to avoid chemical control measures in rice to promote natural control mechanisms. Due to economic constraints pesticides are not easily available in Sierra Leone, a situation which helps the promotion of sustainable pest control methods. Pesticide legislation when developed should be enforced to protect market demands from unscrupulous suppliers.

There has been an attempt to initiate IPM of the armyworm S. exempta in Sierra Leone (James, 1989). Fifteen pheromone trapping stations were installed to detect the presence and monitor the movements of the moths. The monitoring system was established largely along the guidelines of the DLCO-EA/EC Regional Armyworm Project, Kenya (Lambert, 1989).

The AVRDC recommendations and locally collected information for IPM of Cylas weevils in sweet potato, have been tried in the field, and passed on to farmers during a training session. The most important aspect of this is early planting, since weevil incidence increases in the dry season. Other techniques include use of weevil-free vines, removal of Ipomea weeds (alternative hosts) and re-ridging to deny weevils access to the developing tubers. Mass trapping of the adult using pheromone traps has been tried in on-station experiments.

LIAISON BETWEEN THE CROP PROTECTION SERVICE EXTENSION WORKERS AND THE FARMERS

Contacts between these two groups have yet to be developed and intensified to encourage exchange of ideas and information and to enhance the design, development and implementation of IPM packages or to modify existing management practices.

CONSTRAINTS

Training and development in the CPS should be boosted and the personnel motivated, e.g. through a revised salary structure and/or incentive schemes, to increase the reliability of the service. The neglect of the armyworm trapping network, for example, is more the fault of a poorly motivated extension service than the lack of other material inputs. Budgetary support for crop protection research at the agricultural research institutes and university should also be increased. The generation of feasible pest management packages, especially for resource-poor farmers, will depend on support at the research level. The poor development and enforcement of pesticide legislation and plant quarantine rules will also continue to threaten the viability of recommended pest management practices. The possible re-activation of the National Crop Protection Co-ordinating Committee in the near future will hopefully lead to a review of all the constraints.

RESULTS

Six years after the initiation of a biological control project the cassava mealybug problem is less serious. The parasitoid (Epidinocarsis lopezi) has dispersed successfully over the mealybug infested zones and survived alternating rainy and dry seasons since its first release in December 1985 (Neuenschwander, 1985). Soon after the releases, mealybug numbers at the outbreak site were 234 and 134/shoot tip in 1986 and 1987, respectively (James and Fofanah, in press). By 1990, the pest had practically disappeared in the outbreak area; and in four other selected fields in the region, peak mealybug numbers of 8–47/shoot tip occurred in association with 2–25% parasitism (E. lopezi)/month (Leigh, 1990). The high parasitism rates occurred during the early phase of the pest's population growth. The parasitoid's success has been aided by successive rainfall, which reduces mealybug numbers, the farmers abstinence from chemical control of Z. variegatus on cassava, and in some fields, the use of mealybug-free planting materials. For green mite control in cassava the establishment of two recently (February 1992) released phytoseiid species is being monitored. Similarly in mango mealybug control, the establishment and dispersal of the parasitoids, *Gyranusoidea tebygi* and *Anagyrus mangicola* (released in December 1991), are also being monitored. The mango mealybug is still contained in the outbreak area, Lunsar.

IPM strategies to control *Z. variegatus* through large-scale demonstrations have just begun in the country with the initial training programmes. In previous years, a few individual farmers' adherence to spot insecticide spraying of early instar nymphs in citrus fields resulted in only short-lived benefits as the pests recolonized the treated areas from neighbouring fields. It is also premature to provide reliable information on the propagated IPM measures for *Cylas* in sweet potato. Experimental mass trapping of the adult weevils with pheromone traps has been unsuccessful, most probably because the pheromone is known to be specific for *C. formicarius* whilst *C. punticolis* is prevalent in Sierra Leone. Sweet potato farmers are conversant with the identity, seasonal occurrence and damage effects of the weevil, but they are yet to adopt the recommended IPM measures.

In rice, the unintended conservation of natural enemies, through farmer adoption of non-chemical control measures, helps to prevent outbreaks of native pests. In the armyworm monitoring system, peak nightly moth catches per station were always below the economic threshold of 20 moths and no larval incidence was confirmed during the trials (James, 1989). The trapping network is now abandoned, but if resuscitated, it will form a strong and reliable foundation for IPM of the migratory pest nationwide; and will provide information to neighbouring countries.

CONCLUSION

The CPS is now more aware of the benefits of IPM and training has been increased within the service and in many farming communities for the implementation of sustainable crop protection measures. The initial contacts between the CPS and external development agencies, and the co-operative professional links created between the CPS (extension) and research workers can still be improved to lead to the design and development of feasible IPM packages. The CPS field staff are keen learners with a proven ability to pass on extension messages. The opportunity for the successful adoption of IPM strategies is greater now than in the pre-1986 era. In addition to the increased awareness and training, Sierra Leone currently has a poor market for pesticides. The farming community is now ready to be convinced about the advantages of IPM. The removal of constraints to IPM development and testing in Sierra Leone will therefore go a long way in ensuring the successful and early adoption of the management practices by farmers.

REFERENCES

ANON. (1971) *Agricultural Statistical Survey of Sierra Leone 1970/71*. Freetown: Central Statistics Office.

ANON. (1991) *Implementation Agreement*. National Biological Control Programme, Sierra Leone.

AZIM, A. (1988) *Insect Pests of Rice and Their Natural Enemies in IVS, North Western Region, Sierra Leone*. Field Document number AG:DP/SIL/85/004.

FAO (1987) *Strengthening of Crop Protection*. Project Document number SIL/85/004. Rome: Food and Agriculture Organization of the United Nations.

FAO (1991a) *Strengthening of Crop Protection, Sierra Leone. Project Findings and Recommendations*. Terminal Report AG:DP/SIL/85/004. Rome: Food and Agriculture Organization of the United Nations.

FAO (1991b) *Emergency Assistance to Combat Outbreak of Grasshoppers*. Project Document number TCP.SIL.0155 (E). Rome: Food and Agriculture Organization of the United Nations.

JAMES, B.D. (1987) The cassava mealybug *Phenacoccus manihoti* Matt-Ferr (Hemiptera: Pseudo-coccidae) in Sierra Leone: A survey. *Tropical Pest Management*, **33**(1): 61–66.

JAMES, B.D. (1989) Spodoptera exempta: *Initiating a Forecasting System*. Field Document 3. AG:DP/SIL/85/004. Rome: Food and Agriculture Organization of the United Nations.

JAMES, B.D. and FOFANAH, M. (in press) Population patterns for *P. manihoti* Matt-Ferr on cassava in Sierra Leone. *Tropical Pest Management*.

LAMBERT, M.R.D. (1989) *Assessment Mission for the Establishment of an Armyworm Forecasting and Monitoring Network in West Africa (The Gambia, Ghana, Guinea, Nigeria and Sierra Leone)*. Technical Report TCP/RAF/6777. Chatham, UK: Natural Resources Institute.

LEIGH, L.E.A. (1990) *Status and Biological Control of Exotic Cassava Pests in the Western Area of Sierra Leone*. B.Sc.(Hons) dissertation. Fourah Bay College, University of Sierra Leone. (unpublished)

NEUENSCHWANDER, P. (1985) *Mealybug Natural Enemies in Sierra Leone and Monitoring of Establishment in Senegal and Guinea Bissau*. Ibadan: International Institute of Tropical Agriculture.

Integrated pest management in The Gambia

S. KEITA

Department of Agricultural Research, Yundum, The Gambia

INTRODUCTION

Pesticides are an important method of controlling diseases, insect pests and weeds, but despite their widespread use, Gambian farmers still experience a significant loss in yield. Pre-harvest crop losses in The Gambia are significantly high, and with the rapid growth of pesticide resistance in insects, pathogens and weeds, these losses are likely to increase. The Department of Agricultural Research is adopting an integrated approach to address these problems.

To implement these new programmes, a range of techniques that can be used by farmers are already available. They include the use of various cultural practices such as crop rotation, multiple cropping, timing of planting and harvesting, habitat management to enhance the populations of natural predators, biological control agents, direct trapping of insects and finally, the careful and timely use of pesticides.

The advantages of IPM strategies for farmers in The Gambia are many:

- cuts production cost by reducing reliance on expensive agrochemicals

- reduces hazards to both humans and the environment

- stabilizes yields by ensuring the survival of natural enemies of major pests.

IPM RESEARCH AND IMPLEMENTATION ACTIVITIES

Recognizing the high costs, and the dangers pesticides cause to both humans and the environment, the Department of Agricultural Research, in collaboration with the Department of Agricultural Services and some NGOs, is adopting an integrated approach to combat pest problems.

Use of resistant varieties

A wide range of crop varieties have been evaluated for their tolerance or resistance to insect pest damage. The use of bristled early millet varieties against blister beetle (*Psalydolytta fusca*) and millet head-miner (*Heliocheilus albipunctella*) is a notable example. Results from this research showed a low infestation by these pests and this was due to the long bristles which protrude from the spikes.

Similarly, there are ongoing research activities on varieties that are either tolerant or resistant to disease pathogens of economic importance. Some rice varieties (ITA 150 and ITA 257) have been identified as being tolerant to rice blast (*Pyricularia oryzae*). Reports show that up to 80% yield loss in rice has been incurred as a result of attack by this pathogen.

Biological control

The population of predators (*Epidinocarsis lopezi*) released to reduce the damage caused by the cassava mealybug (*Phenacoccus manihoti*) is still being monitored. However, reports received from what used to be highly infested areas show that the population of this pest is on the decline.

Another potential area for biological control that was explored, but was not successful due to some technical problems, was the control of *Raghuva albipunctella* through the release of reared *Bracon hebetor* parasites.

Weed management

Different management strategies are currently being studied to reduce losses due to weeds on farmland. The most important weed species is *Striga* in cereals. Different methods have been adopted, including the use of tolerant sorghum varieties (ICSV 1006, ICSV 1001 and ICSV 1002) against *Striga* infestation.

Another method of controlling *Striga* that was studied was within row intercropping of groundnuts and sorghum. This system however, has some drawbacks in that it is not compatible with the draught animal-based cropping systems of The Gambia with regard to the harvesting and seeding of the crops.

Invariably, the effects of intercropping maize and cowpea was found to reduce the *Striga* population significantly. It was observed that the cowpea variety TN 88-63 cast a dense shade around the base of the maize plants such that *Striga* seeds did not attain the optimum temperature required for germination.

Other research activities on *Striga* included the use of *Cassia obtusifolia* as a biocontrol agent and the effects of duration of competition of *Cassia* on maize yields and *Striga* population. Results from these trials have yet to be conclusive.

Farmers graze and kraal cattle *in situ* in the off-season. This appears to be an effective control measure for controlling the *Striga* population. Trampling on the weed and the high deposit of urine and dung is observed to reduce subsequent seed germination.

CONCLUSION

Trying to eradicate pests completely from field crops has resulted in the development of pest resistance to chemical pesticides, creation of secondary pests, cases of human poisoning, environmental contamination, to cite a few examples.

IPM is the most viable alternative for Gambian farmers. An IPM strategy is less costly and it attempts to manage pest populations at a level that does not cause economic damage.

Integrated pest management in the Côte d'Ivoire

N. COULIBALY
Institut des Forêts, Abidjan 01, Côte d'Ivoire

INTRODUCTION

From the earliest days of its independence, the Côte d'Ivoire has targeted agriculture in its development programme. The authorities set up research and dissemination organizations which enabled the country to achieve good results for both food crops and export crops. Research has facilitated the farmers' task of applying integrated control against the main pests for each crop, insects, nematodes, mites, diseases, rodents, weeds etc. Such control usually involves a combination of control methods, for example, cultural, genetic and biological, which are already available. The main emphasis in the Côte d'Ivoire has been on genetic, biological and cultural control, so as to reduce or eliminate the use of pesticides.

With scientific, technical and financial support from France, agricultural research stations were set up after independence for each of the main crops. These included research stations for: coffee, cocoa and other stimulant plants; rubber; oil seed crops; fruits and citrus; forest species; cotton and other textiles; food crops; and sugar-producing plants.

Each institute had two main tasks:

(a) to develop high-yielding varieties which yield good quality crops;

(b) cultivation, protection, harvesting, storage and post-harvest processing techniques appropriate to the farmers' means, while preserving the environment.

Because of the scale of damage and losses caused by pests, research was directed towards:

(a) pest identification, information on the main pests and their economic effects;

(b) the identification of useful organisms, the part which they play in the control of pests, and their relationship with the biotic and abiotic environments.

In parallel with this work, effective chemicals were selected, along with tolerant or resistant crop varieties; biological control was also tested. The effect of cultural techniques as methods of controlling infestation were also measured.

Following the research, a number of recommendations were made for appropriate control for many pests. These involved the combined use of tolerant varieties, cultural techniques, chemicals etc.

EXAMPLES OF CONTROL METHODS

Cocoa

Mirids:

- systematic disposal of wastes as these are egg-laying sites and, therefore, a source of infestation
- limiting the number of insecticide applications to four a year
- use of tolerant cultivars.

Coffee

Bean borer (*Hypothenemus hampei*):

- prophylactic harvesting in February/March to eliminate the remaining beans, which represent a source of infestation
- use of chemicals above a specific level of damage (5% of fruits bored)
- number of annual applications of chemicals limited to two.

Banana (payo and plantain)

Nematodes:

- application of nematicides as required
- cultivation technique whereby the soil is cleaned after several years of cultivation.

Cotton

Parasite complex:

- application of insecticide on demand
- use of tolerant varieties

BIOLOGICAL CONTROL METHODS

Most pests have natural enemies: insects, viruses, fungi, spiders etc. Surveys of these natural enemies have been made for the main pests of all the crops investigated. Observations show that under favourable conditions, some of these organisms effectively control their hosts. For example:

- *Porops nasuta* on *Hypothenemus hampei* (90–95% attacked)
- *Apanteles* sp. on *Chilo* spp. (80–95% attacked)
- Virus on *Helicoverpa armigera* (90–100% attacked)
- *Trichogramma* sp. on *Earias* spp. (90–100% attacked).

There have also been several successful attempts to develop biological control methods, for example, virus control of cotton-bush *Helicoverpa* by IAESSA and *Trichogramma* on *Earias biplaga* by ACC-IAEFOR. However, these methods were not developed further because of a number of constraints: instability of the virus preparations to ultra-violet light, the high costs of the mass production of *Trichogramma* etc. These constraints partly explain why biological control, in practice, is often limited.

Control of cassava pests

The Plant Protection Department, which is responsible for implementing plant protection policy and legislation, acts when the need arises to organize plant protection treatments for crops which are not satisfactorily covered by an extension service. The Department has been responsible since 1986 for the biological control project against cassava pests initiated and partly financed by IITA which provides technical support. This is probably the only biological control project in the Côte d'Ivoire which benefits from external financial support.

The project has concentrated on three main areas:

(a) surveys to discover the dynamics of cochinellid populations;

(b) evaluation of levels of infestation and levels of damage;

(c) releases of natural enemies and evaluation of their impact.

Several technicians have benefited from training provided by the Plant Protection Department as part of the project. However, only one technician is presently working on the project.

As the project has progressed, several shortcomings have become clear: the research dimension was not fully considered at the outset; farmers were not involved in the project; and any effects are difficult to assess since there was no diagnostic work before the releases. Furthermore, releases were made during the rainy season.

There are a number of reasons why biological control methods are not as widespread as they might be. First, the investigation work (knowledge of the pest, its natural enemies, host-parasite relationships, interactions with the biotic and abiotic environments etc) is long and costly. So too is development of the method (rearing of parasitoids, release techniques, release times etc.). Second, there is a need for skilled and experienced personnel, as well as for substantial infrastructure and equipment. Third, financing has, so far, come out of the limited budgets of the research institutes. This has meant that research has been underfunded, and this has slowed down progress on finding usable results for immediate application.

Against this background, and given the scale of crop losses it is not surprising that so much emphasis has been placed on pesticides. Also much research work in the Côte d'Ivoire is financially supported by chemical companies.

Prospects

Because the disadvantages of pesticides have become clearer in recent years, more attention is increasingly being given to biological control used in conjunction with cultural and genetic techniques, although the latter is both costly and slow to implement.

Control of pests is increasingly being approached in a multidisciplinary way, involving, for example, genetics, cultural techniques, entomology, plant pathology and weed control.

A restructuring at institutional level (described below) should make it possible to achieve the use of integrated control for most crops, with little or no use of chemicals.

A restructuring of the production system and research institutes is underway, and, once completed, will give both the research institutes and agriculture generally a new impetus. In the long term, this will be reflected in a withdrawal by the government in favour of farmers working through social and professional organizations, alongside the private sector.

As part of the restructuring process, two new institutes, partly state-owned, were set up in 1992.

(a) IDEFOR (Institute of Forestry) which includes the Department of Coffee, Cocoa and other stimulant plants; the Department of Latex Plants; the Department of Oil-producing Plants; the Department of Fruits and Citrus Fruits; the Department of Forestry.

(b) IDESSA (Institute of Savannas), which was set up in 1982, but with the status of a national public establishment.

CONCLUSION

The Côte d'Ivoire owes its economic success to agriculture. Research has led to the adoption of rational methods of production, including IPM. Although chemicals are used on a massive scale for cash crops as opposed to subsistence agriculture, there has been considerable effort to persuade farmers to use other pest control methods, in particular biological control.

Several constraints such as high costs, the length of time needed to carry out research, a shortage of experienced workers etc, have meant that the hoped-for results have not been achieved yet. However, given that the disadvantages of chemicals are becoming clearer, and in view of the structural improvements at institutional level, increasing efforts applied to both genetic and biological control should be more successful in future.

Discussion

The speaker from Sierra Leone was asked to clarify a remark made during his presentation about the need to refine existing methods of IPM. He responded by explaining that in some cases existing IPM methods could be improved or better adapted to the local situation, and that, more generally, better co-ordination of various methods should be pursued, taking into account economic, political, health and environmental factors.

S. Keita from The Gambia was asked to explain the conditions surrounding the discontinuation in 1988 of the national IPM programme which had been achieving considerable success up to that time. The speaker explained that this successful IPM government service had been disbanded due to economic factors. Although international donors did step in to maintain some of the services, declining trends in agricultural production since 1988 stress the need for a strong national crop protection service.

Integrated pest management in Cameroon

M. TCHUANYO

Institut de la Recherche Agronomique, Ekona, Buea, Cameroon

INTRODUCTION

Cassava is the main root crop in Cameroon, with an output of more than 1 million t/year. Like most cultivated plants, cassava is prone to attack by many diseases and insects, leading to considerable damage and loss of yields and consequently, income.

The main pests which attack cassava in Cameroon are:
- elegant grasshopper (*Zonocerus variegatus*)
- cassava mealybug (*Phenacoccus manihoti*)
- cassava green mite (*Mononychellus tanajoa*)

Over a number of years, research has been undertaken at the Agronomic Research Institute of Cameroon, as part of the National Programme for the Improvement of Tuber and Root Crops, with a view to controlling the cassava mealybug and the cassava green mite. Research has concentrated on two main areas:

(a) the development of tolerant varieties;

(b) biological control of the cassava mealybug and the green mite.

CURRENT STATE OF RESEARCH

National surveys carried out in the early 1980s indicated the main constraints limiting the production of cassava as follows:

(a) diseases: mosaic virus, bacteriosis (*Xanthomonas campestris* pv. *manihotis*) and black spot (*Colletotrichum gloeosporioides* f. sp. *manihotis*);

(b) insects and mites (the elegant grasshopper, cassava mealybug and cassava green mite).

Development of resistant/tolerant varieties

The selection of clones which were resistant/tolerant to the main diseases and the green mite began in 1982. Using the method developed by IITA, the source population was subjected to a number of stages (5–6) under different selection pressures for resistance/tolerance to diseases and the green mite. There was no evidence that the mealybug was present in Cameroon when this selection programme started. This pest was, therefore, not taken into account in the selection process.

Biological control of the cassava mealybug and the cassava green mite

The cassava green mite was first observed in Cameroon in 1981 in the Garoua-Boulai area. The mealybug was observed in 1985 by Hammond and Tchuanyo in the frontier region between Cameroon and Nigeria.

More extensive surveys during 1985 and 1986 delimited the extent to which the two pests were present. The green mite was found in all areas where cassava was cultivated. The mealybug is widespread in the southwestern province (areas close to the frontier with Nigeria) and the southern province (frontier areas bordering Gabon and Congo).

Pre-release test for the parasitoid (*E. lopezi*) of the cassava mealybug

Between December 1988 and January 1989 a test to evaluate the extent of the mealybug outbreak and to investigate the possible presence of the parasitoid originating in neighbouring countries, revealed that *Epidinocarsis lopezi* was present in the regions bordering Nigeria and Gabon. This survey was carried out just after the authorities authorized the introduction of the parasitoid into Cameroon.

100

In January and February 1991 a further pre-release survey on the two pests was carried out throughout Cameroon. Particular emphasis was placed on the green mite because of the severity of the damage it causes. These data are presently being analysed.

To summarize: the biological control programme in Cameroon operates under the National Programme for the Improvement of Tuber and Root Crops at the Agronomic Research Institute of the Ministry of Higher Education, Data Processing and Scientific Research. In the absence of any one organization responsible for biological control, the delay in the introduction of natural enemies, as well as the lack of financial support, have restricted IPM work in Cameroon. Laboratory equipment provided as part of the FAO/TCP project in 1986 and the FAO/RAF project in 1988 has facilitated some IPM work. However, transport and finance limitations are serious handicaps to sustained and integrated work in the field.

Integrated Pest Management in Chad

A.M. NGARE

Direction de la Protection des Vegetaux et du Conditionnement, Chad

INTRODUCTION

In Chad arable land is estimated at 320 000 ha (out of a total land area of about 13 million ha). In the south of the country the dry season lasts from December to April, while the rainy season is from May to November. In the north the dry season lasts for 7–8 months. About 77.4% of the population is involved in agriculture.

The most important crops are: millet cultivated on approximately 950 000 ha, giving yields of 573 kg/ha; cassava on 70 000 ha yielding 4385 kg/ha; rice paddy on 20 000 ha yielding 1250 kg/ha; groundnuts on 134 000 ha yielding 781 kg/ha; and seed cotton on 160 000 ha yielding 688 kg/ha.

CROP PROTECTION

From 1985 to 1987 USAID funded an IPM project with the following objectives:

- identification of pests of millet and sorghum

- determination of the importance of these pests

- studies of natural factors which limit pest activity

- implementation of IPM pilot studies in four areas.

The project was managed by FAO and carried out by CILSS and the Department of Plant Protection in Chad.

There is no national biological control unit but there is a Department of Plant Protection (Direction de la Protection des Végétaux). Two members of the Department were trained in biocontrol in 1988 with the intention that they undertake country-wide surveys (southern provinces) and maintain feedback. In addition they were to form a nucleus for the initiation of biological control activities in Chad.

Cassava mealybug (*Phenacoccus manihoti*) had earlier been sighted along the Lake Chad basin by one of the IITA Biological Control Programme scientists but the extent of infestation could not be confirmed and therefore, the introduction and release of natural enemies could not take place.

As well as the cassava mealybug infestation in Chad, an exotic scale pest (*Aonidela aurentalis*) seriously infests most neem trees around the Lake Chad basin affecting countries such as Cameroon, Nigeria and Niger. Neem trees are an important source of shade, fuel wood, hedges etc., and are readily adaptable to arid areas such as found in Chad. Initiating biocontrol, therefore, will not only benefit cassava plants but neem trees as well. Besides, Chad is heavily dependent on the use of pesticides for control and is aware of the detrimental consequences of using chemicals.

Officials of the Ministry of Agriculture have shown keen interest in biological control, but as yet little has been done in practice. There is no infrastructure to support a biological control programme. There are no suitable buildings to accommodate a programme and transport and finance are not available. The country has been battered both by natural disasters, such as drought and locusts, and war with a neighbouring country, and has few resources to support activities such as biocontrol.

Should the above problems be solved, the development of biological control in Chad could proceed with few difficulties.

Chad is a member of the Programme Africain de Lutte Biologique/IITA — UNDP/FAO RAF/87/142, and has benefitted from a training course organized by the programme.

Integrated pest management in the Congo

G. BANI and J. BANTSIMBA

Office de la Recherche Scientifique Outre-Mer (ORSTOM), DGRST, B.P. 181 Brazzaville, Congo

The chemical arsenal of conventional insecticides is now highly diversified. Chemicals are very effective, although they are also sometimes hazardous, to judge from toxic side-effects. They have been referred to as ecological drugs. Nevertheless, humanity owes much to pesticides and, despite undesirable side-effects, their efficiency in pest control must also be recognized.

In the tropical environment, however, there must be certain restrictions on the large-scale use of pesticides, for several reasons:

(a) the high cost which cannot be supported by subsistence farmers, and which is not economic for all cash crop cultivation;

(b) the fact that both useful and harmful insects are destroyed, with the former frequently more sensitive than the latter because of their more exposed habitat or behaviour;

(c) the risk of intoxication in those responsible for their handling, because of a lack of training or inadequate handling equipment;

(d) the increasingly frequent cross-resistance of some insects to whole groups of products following the poor management of chemical control;

(e) the concentration of residues and the problem of pollution;

(f) the lack of understanding of the developmental dynamics of pest problems, particularly by farmers;

(g) insufficient knowledge about pest life styles and their most vulnerable stages;

(h) finally, underestimation of such factors as distance, equipment, maintenance and economic viability.

These factors call for pest control methods to be managed more efficiently. This will enable more imaginative, less conventional and more appropriate methods of control to be put in place.

One possible solution is IPM which, according to the FAO, is a 'system for controlling pest populations which, given the particular environment and the dynamics of the populations in question, uses appropriate methods in as compatible a manner as possible and maintains pest populations at a level where they do not cause economic damage'.

Against this background, biological control is the appropriate method for developing countries which lack the financial resources to purchase expensive pesticides, to train personnel in plant protection or to set up a system for monitoring environmental effects of pesticides.

At present, IPM in the Congo is directed towards: the cassava mealybug, *Phenacoccus manihoti*; the cassava green mite, *Mononychellus tanajoa*; the mango mealybug, *Rastrococcus invadens*; and the elegant grasshopper, *Zonocerus variegatus*.

Biological control of *P. manihoti* began in 1982 with the introduction and release of *Epidinocarsis lopezi*, and the coccinellids *Hyperaspis raynevali*, *H. jucunda* and *Diomus* spp.

Epidinocarsis lopezi acclimatized well. Its biology has been investigated. In the field, the development of populations of *E. lopezi* and the development of populations of its regulating agents have been studied in relation to climatic factors. The results obtained showed the valuable biological qualities of *E. lopezi* in biological control, but its inability to control cassava mealybug populations alone, from a single release.

Nevertheless, recent unpublished work tends to show that *E. lopezi* is effective in biological control when the parasitoid is released several times. Several hypotheses for strengthening the effect of *E. lopezi* in the Congo are currently envisaged. These include the introduction of new parasitoid species and exotic predators.

Research is also being carried out on host plant/pest relationships, for example, on the mealybug's biotic potential for different varieties of cassava under various conditions of soil fertility. The object is to determine the effect of fertilizer use on the biotic potential of the mealybug to enable an agricultural control strategy to be implemented.

At present biological control of the cassava green mite is only at the stage of listing its local, natural enemies. Some predators have already been identified. These are:

Eusseius fustis and *Typhlodromalus*, Phytoseidae;
Sthetorus spp., Coccinellidae
Oligota spp., Staphylinidae
Cecidomyidae and Thysanoptera.

Their efficiency against green mite outbreaks is considered to be negligible.

The presence of *Gyranusoidea tebygi*, introduced into Togo to control *Rastrococcus invadens* was reported in the Congo in 1989. Since then studies of population dynamics have been made. The level of pest populations fell greatly after the accidental introduction of this hymenopteran.

As far as the elegant grasshopper, *Zonocerus variegatus*, is concerned, research shows that a change in cultivation techniques, together with keeping the pest population at a certain threshold, will protect cassava from early attacks, which are the most harmful. Work is currently underway to find a biological means of controlling the level of populations of *Z. variegatus*. This involves searching for parasites (insects, fungi or bacteria) of the eggs and larval forms of *Z. variegatus*.

PROSPECTS FOR PROGRESS

There is considerable potential for the development of IPM in the Congo. New programmes have been started in a variety of fields, for example:
* control of the banana tree weevil, *Cosmopolites sordidus*;
* control of the sugar-cane borer, *Eldana saccharina*;
* control of citrus fruit scale insects;
* control of *Chromolaena odorata*.

Other potential research topics include:
* biological control of *Mononychellus tanajoa* through the introduction of predatory mites (Phytoseidae),
* the biological control of *Alcyrodes* through the use of the pathogenic fungus *Ashersonia alcyrodis*.

IPMWG DISCUSSION PAPER

Strategic issues in West African IPM*

A. KREMER
Overseas Development Administration, London, UK

INTRODUCTION

This workshop brings together donors and West African institutions to discuss the implementation of IPM. The donors have come for a specific purpose, i.e. to find out in which direction IPM must move to produce cost-effective and sustainable improvements in farmers' welfare. West African institutions, concerned with plant protection, have an equally well-defined reason for attending; their ability to work on plant protection in the future will depend in part on their success in persuading donors that pest management technologies can help farmers. One can therefore, see this workshop not as a scientific conference but as a 'board meeting' where the participants conduct a dialogue intended to identify the most profitable business strategy.

This paper raises some of the issues that should concern such a board meeting. It poses questions on the objectives of IPM and the way in which IPM projects are organized, on the role of donors and the nature of their dialogue with recipient institutions. It poses questions—but it does not answer them. That is the job of the workshop itself.

RECENT DEVELOPMENTS IN WEST AFRICAN IPM

The definition adopted by the IMPWG, 'IPM concerns the farmer's best mix of control tactics in respect of yields, profits and safety of alternatives' indicates the range of what might be acceptable to farmers and qualify as IPM. By its nature IPM is diffuse and difficult to summarize but a brief résumé of recent developments, based upon the country reports submitted by workshop participants, is necessary for those unfamiliar with the subject. For this purpose one may divide the country reports into two regions, a humid and sub-humid sub-region (Central African Republic, Cameroon, Nigeria, Benin, Togo, Ghana, Sierra Leone, Guinea and Guinea-Bissau) and a Soudano-Sahelian sub-region (Niger, Burkina Faso, Senegal, Chad and The Gambia).

The humid and sub-humid sub-region

IPM in the humid and sub-humid sub-region was strongly influenced by the development of a 'classical biological control' programme against the cassava mealybug (*Phenacoccus manihoti*), an introduced pest of cassava, which became established in Africa in the 1970s. Its arrival set off a chain of events that followed a similar pattern in several countries. Expatriate experts would identify it and a donor would then agree to fund postgraduate training for local scientists and a survey of the mealybug and its predators. IITA would then be employed to assist with the release of the parasitic wasp *Epidinocarsis lopezi* and ex-post evaluations would give varying accounts of its effectiveness.

The organization of IPM in this sub-region appears to have been determined by this classical biological control model. For most countries the initiative for non-chemical pest control came from outside and led to the creation of separate budgets for units concerned with biological control within plant protection services, for a specific period. According to the country reports, farmer involvement has been minimal, being ruled out, in the words of one participant, "by the specialized nature of these activities". The role of extension services has, therefore, been small.

Notwithstanding, country reports from the Southern zone make many references to improved varieties and timely cultivation as possible means of reducing pest damage. They note, however, that these technologies have not been readily adopted. They attribute this to a lack of funding for research and the weakness of national extension services.

The Sudano-Sahelian sub-region

In 1978, an IPM project, funded by USAID, CILSS and the FAO, was established (as a complement to USAID's regional food crop protection project) with the mandate to identify, test and extend

---

* This paper was commissioned by the workshop organizers as a resource paper and circulated to participants in advance to stimulate debate during the workshop on the various issues raised.

IPM techniques throughout the western Soudano-Sahelian region. Due to administrative and management problems related to the tripartite arrangement, the programme was terminated in 1987, at a time when Sahelian research capacity was beginning to develop but before much IPM technology could be adequately developed or extended to farmers. Although the IPM project is largely considered a failure for these reasons, it should be recognized that during the life of the project numerous crop protection services were established or strengthened, and a core group of crop protection professionals, many of whom are presently employed by national crop protection services, were trained.

One of the project's first activities was to establish a list of target pests. This included the millet head-miner (*Heliocheilus albipunctella*), meloid beetles, the millet stem borer (*Coniesta ignefusalis*), grasshoppers, locusts, mildew, smut and *Striga hermonthica*. Various possible control methods were identified. It was suggested that the wasp *Bracon hebetor* might control *H. albipunctella*, that *Smicronyx* spp. might control *S. hermonthica*, that insecticide dusts were effective against meloid beetles. The project closed, however, before confirming or refuting these suggestions and many others.

Since 1986 investment in non-chemical pest control in the Sahel has been insignificant compared with expenditure on insecticide campaigns against grasshoppers. National plant protection services have concentrated upon the supply of free sprayers and chemicals to farmers. This is not to say that non-chemical control has been forgotten. Fungal pathogens have been tested against grasshoppers in Mali and Niger, grasshopper egg-pod digging has been encouraged in Mali and Burkina Faso has taken up the study of the effect of *Smicronyx* spp. upon *S. hermonthica*. Also, village or farmer brigades for 'self help' in crop protection have been established throughout many of the Sahelian nations.

Various crop protection initiatives have also been undertaken in the region through the support of foreign donors. These include:

  FAO/UNDP crop protection strengthening programme in Mali
  FAO/Dutch crop protection strengthening programme in Chad
  CIDA crop protection research and extension programme in Burkina Faso
  CIDA long-term support to the CPS in Niger
  GTZ long-term support to the CPS in Niger
  Dutch support to the regional plant protection training centre, DFPV, Niamey, Niger.

ISSUES AND QUESTIONS CONCERNING EFFECTIVE IPM IMPLEMENTATION IN WEST AFRICA

At the risk of over-generalization, one can make the following observations about the current state of IPM in West Africa:

(a) many potential IPM technologies have been identified but few have been tested for their appropriateness to farmers' needs;

(b) some West African plant protection institutions have acquired experience in classical biological control, others have acquired experience in the management of pesticide campaigns; few have experience, however, of relating new pest control techniques to farmers' needs.

These observations raise a number of important questions about the direction of IPM policy in West Africa. These issues will be expanded below.
• What is good IPM? With so many ideas to follow up, how do we decide which are the most promising?
• Are research, plant protection and extension services organized in a way that is suited to IPM work?
• What is the right kind of donor funding for IPM?

What is good IPM?

With so many potential technologies identified but untested, there is a real need for agreement on what is the defining feature of a good IPM technology. 'Integrated pest management' is a vague phrase and, as with other vague phrases, its definition varies according to the preoccupations of the speaker. A donor may define IPM in terms of a reduction or elimination of the use of chemical pesticides. A biologist may define it in terms of the use of interactions between crop, pest and predator, an interpretation implicit in the French phrase 'lutte biologique intégrée'.

For most people, IPM is defined, not in terms of what it is, but in terms of what it is not: the uncritical use of chemical pesticides. The IPM approach gained popularity in Southeast Asia and Central America because excessive pesticide use in these zones had led farmers into economic difficulties through pesticide resistance and the elimination of pests' natural enemies. According to Kiss and Meerman (1991) this process provides the rationale for IPM. In West Africa, however, pesticide use is relatively low (Table 1), cotton and horticulture are possible exceptions, and the 'pesticide treadmill' is therefore, unknown to almost all small farmers. Indeed, the application of pesticides to cereals in the Soudano-Sahelian countries is restricted largely to chemicals provided free of charge by foreign donors.

Table 1. Insecticide use, 1979–81 (kg/ha arable land/year)

Country	Insecticide use (kg/ha/year)
West Africa	
Niger	0.07
Senegal	negligible
Sierra Leone	negligible
Central America	
El Salvador	1.37
Guatemala	0.38
Mexico	0.21
Southeast Asia	
Indonesia	0.14
Philippines	0.16
Thailand	0.29

FAO Production Yearbook (1989).

If IPM is of relevance to these West African small farmers, it is not because pesticides have recently become ineffective but because they have always been too expensive. It can therefore be argued that the *defining characteristic* of IPM in West Africa should not be an enthusiasm for non-chemical technologies but a requirement that the process of technology generation and diffusion should be driven by farmers' economic circumstances. Hence, the particular relevance to the region of the Consultants' Report's description of IPM as a system wherein the farmer chooses the combination of techniques most appropriate to his or her economic environment (NRI, 1991).

The current identification of IPM with non-chemical, biological control has led to some paradoxical situations in which the breadth of crop protection options has been reduced. In the Sahel, for example, it is easier to secure funding for research into fungal pathogens of grasshoppers than into synthetic pyrethroids. Although more expensive, less effective and more difficult to store, fungal pathogens carry the label 'IPM', so they are in fashion.

In fact, it can be argued that a reduced use of chemicals and an increased use of biological interactions are only incidental to good pest management. This may mean increased or decreased insecticide use, and pest/predator interactions may have some or no place in the farmer's choice of technology. It might therefore, be useful for workshop participants to discuss what constitutes good IPM. Is it reduced pesticide use and increased use of biological interactions or decision-making by farmers? The answer to this question will open many other avenues of discussion.

With so many ideas to follow up, how do we decide which are the most promising?

The country papers submitted by workshop participants mentioned many possibilities for pest control that have been described as promising by researchers without being evaluated for their effectiveness or appropriateness to farmers' needs. A small selection is listed in Table 2.

A frequent comment in the country papers was that these 'techniques-in-waiting' could be brought to implementation if more funds were available for research, pilot programmes and

Table 2. Pest control techniques – identified but of uncertain value

Country	Pest	Technique/natural enemy
Burkina Faso	millet head-miner	*Bracon hebetor*
	Striga	insects
Cameroon	cotton insects	plant growth regulators
	cotton insects	threshold spraying
Congo	cassava mealybug	fertilizer
	Z. variegatus	cultural control
	M. tanajoa	predatory mites
The Gambia	meloids	smoke
	meloids	baobab fruit
Mali	grasshoppers	egg-pod digging
	grasshoppers	bare field edges
	grasshoppers	insecticide baits
Niger	millet stem borer	resistant varieties
Senegal	*Striga*	weeding
	Striga	*C. occidentalis*
	Striga	*Smicronyx* spp.
	millet stem borer	scorching stems
	millet head-miner	deep ploughing
	millet head-miner	nitrogen fertilizer
Togo	cotton insects	microbes
Nigeria	cassava mealybug	classical biological control
		resistant varieties
		cultural controls
Sierra Leone	African armywork	moth trapping
	cassava green mite	phytoseiid species
	cassava mealybug	*Epidinocarsis lopezi*
	Cylas weevils	timely planting
		Ipomea weeding
		re-ridging
		pheromone trapping
	mango mealybug	*Gyranusoidea tebygi*
		Anagyrus mangicola
	S. exempta	pheromone trapping
	Z. variegatus	timely weeding
		underbrushing
		egg-pod destruction

extension. Donors are therefore, frequently faced with requests for funding for a specific line of research.

It is a fact that the dialogue between researchers and donors is littered with misunderstandings. Researchers frequently describe the decisions of funding agencies as arbitrary and unjust. Funders frequently describe researchers' project proposals as extravagantly over-optimistic. These stereotypes will certainly be familiar to all the participants at this workshop.

The problem is that we lack objective criteria in terms of which to discuss the relevance of a line of IPM research and development. One is thus reduced to describing a technology as 'promising', 'extremely promising' and so on, terms which indicate nothing except the researcher's enthusiasm. Donors and researchers therefore, need a new set of terms, a new set of measures, with which to continue their dialogue upon IPM.

It could therefore be argued that farmers themselves should provide the yardstick in terms of which the IPM technique is described and discussed. Before a line of research is approved it would be possible, by means of rapid surveys, to establish farmers' perception of the pest, farmers' ranking of the pest relative to other pests and non-pest problems, farmers' expenditure upon agriculture relative to the cost of the new technique and so on.

This, therefore, provides another possible question for discussion in the workshop: How could farmers' priorities strengthen the IPM project appraisal dialogue between researchers and donors? By addressing this issue one could avoid the fate of one recent West African research programme that spent US$ 2.5 million trying to solve a pest problem that did not worry farmers.

Does IPM require a new approach to research?

In the classical model of research and development the effectiveness of a technique is initially proven under controlled experimental conditions. The technique is then promoted in the real world, where conditions are so variable that it is hard to measure its effectiveness by means of experimental methods.

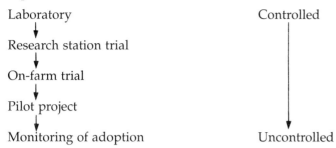

The IPM approach, however, reduces the relevance of controlled experiments. The very essence of a controlled experiment is that it models a complex situation with many variables as a simple situation with few variables. The very essence of the IPM approach, however, is that it admits the relevance of a myriad of ecological and socio-economic variables.

It is therefore, possible that pilot projects are a more appropriate testing-ground for IPM technologies than research station trials and on-farm trials. Workshop participants may wish to discuss to what extent this is the case. If pilot projects are to be the test-beds for IPM, then other questions follow. How is a technology's success or failure in a pilot project to be measured? In terms of its acceptability to farmers? If so, does this imply that all pilot projects require a contribution from social scientists such as economists and sociologists?

Will institutions have to re-organize themselves for IPM?

Research, extension and plant protection services were set up so as to be suited to the classical model of research and extension described above. It does not necessarily follow, however, that these structures will remain suited to the IPM approach. For example:

(a) IPM makes the assessment of farmers' priorities more important; plant protection and extension projects may therefore, need to recruit social scientists or, better still, familiarize technical scientists with the social scientist's approach;

(b) the multiplication of pilot projects will affect the demarcation line that divides the respon-sibilities of the plant protection service from those of the research service; in Mali, for example, the respective roles of the SNPV and the IER in the management of pilot projects remain to be determined at ministerial level;

(c) the IPM approach specifies that pest management is an integrated part of crop management (Teng, 1985) and should therefore, not be treated distinctly: seen in this light, it would seem no more logical to have a separate crop protection service than to have a separate 'fertilization service' or 'cultivation service';

(d) researchers tend to specialize in pests of a particular taxonomic group: if IPM really is to deal with the multiplicity of ecological and socio-economic interactions, as its proponents wish, then specialization by zone or, failing that, by crop would be more logical;

(e) scientific qualifications such as doctorates have until now conferred the rank of 'pest control expert' upon the holder. In the future, however, farmers' perceptions may become the principal source of authority upon what is and what is not viable. This could have long-term consequences for the status of scientific expertise within projects and pest control institutions.

If IPM is to be defined in terms of farmer participation it will constitute a radical departure from existing approaches to pest control. Workshop participants may therefore, want to discuss whether such a radical change requires corresponding changes in institutional organization.

Does IPM need a new kind of project?

One should not avoid the issue: it is difficult to apply traditional principles of financial management to IPM projects. While financial control usually requires pre-stated objectives, quantifiable indicators and precise budgets and timetables for the achievement of objectives, IPM projects require a certain amount of flexibility in order to adapt project objectives and budgets as necessary. IPM projects also put more emphasis on qualitative indicators (such as farmer behaviour) and may often require long-term funding than the usual five year budget maximum.

These contradictions between the requirements of iterative pilot projects and linear financial control can render relations between IPM researchers and their funding agencies problematic. As the requirements of the two parties differ, there may be no solution to this problem, but it would surely be valuable for workshop participants on both sides to discuss it explicitly.

CONCLUSIONS

This paper has argued that a workshop upon the implementation of IPM raises a number of vital strategic questions. Failure to address these questions directly and explicitly could lead to inconsistencies or errors in IPM policy. These questions are as follows:

Definitions
What is good IPM? Is it defined in terms of reduced pesticide use and increased use of biological interactions or in terms of farmer participation?

Project appraisal
How could the project appraisal dialogue between donors and researchers be made more meaningful? Could the value of a line of IPM research be described in terms of farmers' priorities?

Research methods
Is the current mix of experimental work and pilot projects the right one? How could the success of a pilot project be measured? Does IPM require social science expertise?

Management
Are our institutions' organizational structures suited to IPM? Can one reconcile donors' need for financial control with researchers' need for flexibility?

These questions are at the root of IPM policy-making in West Africa. It is to be hoped that workshop participants will find time to address them directly.

REFERENCES

KISS, A. and MEERMAN, F. (eds) (1991) *Integrated Pest Management and African Agriculture. World Bank Technical Paper 142.* Washington D.C.: World Bank.

NRI (1991) *A Synopsis of Integrated Pest Management in Developing Countries in the Tropics. Synthesis report commissioned by the Integrated Pest Management Working Group.* Chatham, UK: Natural Resources Institute.

TENG P.S. (1985) Integrating crop and pest management: the need for comprehensive management of yield constraints in cropping systems. *Journal of Plant Protection in the Tropics,* **2** (1): 15–26.

SUB-REGIONAL WORKING GROUPS' REPORTS

Constraints to the implementation of IPM (not in order of priority)

SUDANO-SAHELIAN SUB-REGION

Political/policy

- Lack of government policy on plant protection, including IPM, in many countries of the region
- Lack of adequate recognition/awareness by policy-makers of IPM as a tool for controlling pests
- Free pesticide donations are sometimes used by policy-makers as a political tool

Institutional

- Lack of adequate national infrastructures for research, training, extension and information dissemination
- Lack of co-ordination between institutions, extension service etc.
- Lack of long-term sustainable plans for the development of plant protection
- Lack of adequately trained personnel, materials, information and appropriate extension support for farmers

Socio-economic

- The advantages of IPM over other methods of control are not fully understood by farmers
- Lack of information on the socio-economic environment of the farmer that could facilitate the introduction of new technologies
- Lack of available inputs for the implementation of IPM at the required time
- Lack of organization at farmer level
- Lack of adequate marketing outlets and attractive prices for farmers' produce as an incentive to IPM
- Lack of farmer involvement in the identification and implementation of research in IPM

Technical

- Lack of adequate technology to implement IPM in the field
- Lack of reliable systems to evaluate the benefits of IPM
- The perceived complexity of IPM constitutes a constraint to simple approaches
- Poorly focussed IPM research

HUMID AND SUB-HUMID SUB-REGION

- Lack of national policy and legislation for IPM
- Lack of political commitment for the sustainable development of IPM
- Limited national budget for IPM
- Lack of communication and co-operation between policy-makers and technicians
- Lack of communication between farmers, extension workers and policy-makers

- Inadequate liaison between international, regional and national institutions
- Proliferation of national organizations without co-ordination
- Lack of adequate infrastructures for research and training in the region
- Lack of clarity in defining national research priorities
- Lack of co-ordination with NGOs
- Unorganized NGO participation
- Poor budgetary support from government
- Unstable government policies

- Little or no farmer involvement in development and implementation of IPM
- Inadequate socio-cultural considerations in educating farmers, and a lack of farmer education in IPM
- Inadequate infrastructure, logistic and financial support for IPM work
- High costs and non-availability of inputs to encourage the adoption of IPM by farmers (as opposed to subsidized chemicals)
- Abuse of chemical control (use of wrong chemicals or use on wrong crops)
- Religious and cultural constraints leading to conservatism in the adoption of new techniques

- Incomplete data on pests
- Ecosystems are not well defined and characterized
- Research information not published
- Insufficient trained technical staff
- Lack of access to genetic resources
- Inadequate facilities

111

Initiatives for overcoming constraints to IPM implementation

SUDANO-SAHELIAN SUB-REGION

HUMID AND SUB-HUMID SUB-REGION

Political/policy

- Produce information brochures designed to convince policy-makers of the advantages of IPM
- Organize a meeting of policy-makers from the different countries concerned to brief them on IPM; this should be done at the ministerial level and funding sought from potential donors
- Donors should insist on the adoption of IPM as a prerequisite to providing funds for specific projects
- Donors should be encouraged to give pesticides only in response to professionally determined needs; blanket donations of pesticides should be avoided

- Well-defined national policy on agricultural development with emphasis on IPM
- Sub-regional phytosanitary legislation
- Decentralization of administrative procedures governing agricultural development
- Initiation of IPM programmes must involve farmers and technicians who would in turn advise policy-makers
- Strengthen and facilitate the collection and dissemination of appropriate IPM information among scientists within and between countries in the sub-region
- The use of the mass media by agricultural specialists to disseminate information on IPM
- National governments must make adequate funds available to support IPM
- National governments must honour obligations to pay counterpart funds for all IPM projects signed with donor agencies
- Technicians must ensure continuation and sustainability of all agricultural development policies

Institutional

- Study the needs of IPM in each country and develop strategic plans at national and regional levels; this should be done through national IPM Working Groups
- Develop training programmes at all levels by taking into consideration the targeted outputs; ideally, this should be institutionalized by setting up a centre for IPM
- Provide solutions for all the other institutional constraints showing lack of proper attention

- IPM research development and implementation should be centrally co-ordinated in each country (including NGO activities)
- Institutional responsibilities should be clearly defined
- Evaluation of regular dialogue between the IARCs and NARCs involved in IPM
- Adequate allocation of funds should be made locally and, where possible, external funding sought for IPM support and sustainability

Socio-economic

- Organize farmers in small groups and run training programmes in the form of on-farm demonstrations to promote IPM. This training should be done through informal adult education using training aids such as videos and brochures, and should place special emphasis on the training of women, given the crucial role they play in agricultural development
- Encourage low-input practices and recommend the timely and free provision of inputs by donors

- Active farmer participation in the initiation of research and implementation of IPM should be encouraged
- Farmers and extension workers should be trained to have an analytical approach to problem management and to appreciate more fully the advantages of IPM
- IPM packages should take into account the socio-cultural constraints and potentials of farmers; socio-economic studies need to precede the development of IPM packages
- Strong government and donor support (i.e. improvement of infrastructure, logistics, financial and human resources) is a prerequisite to IPM development

SUDANO-SAHELIAN SUB-REGION	HUMID AND SUB-HUMID SUB-REGION

SUDANO-SAHELIAN SUB-REGION

HUMID AND SUB-HUMID SUB-REGION

- Develop simple, practical, cost-effective IPM packages to encourage farmer adoption
- Government subsidies and donor assistance on chemicals be 'transferred' to encourage IPM development and implementation

Technical

- IPM is an ongoing process and should be continuously supported by research efforts
- The many small projects existing in countries should be integrated into national agricultural programmes with full participation of nationals
- Research needs should be identified and research plans developed to address specific problems; once this is done full project proposals should be prepared for submission to donors

- Pest diagnosis (extensive and intensive)
- Ecosystem analysis (environment, plant, pest, natural enemies, interactions)
- Development and implementation of interventions (biocontrol, resistance breeding, cultural control, legislation etc.)
- Improve information services, including improved access to printed literature and an African journal of IPM
- Increase research experience through better liaison and communication
- Introduce a reward system for productive scientists (e.g. those who publish)
- Training at all levels

Table 1. Sub-regional crop/cropping systems and pest priorities

Crops/crop systems*	Pests*
Sudano-Sahelian sub-region[†]	
Sorghum-millet-maize-based cropping system	*Striga*
	grasshoppers and locusts
	storage pests
Cowpea	*Striga*
Rice	pod sucking bugs/pod borers
	stem borers
	weeds
	rice blast
Vegetables	nematodes
	lepidopterous pests
	whiteflies
Cotton	bollworms
	bacterial blight
	cotton stainers
Groundnuts	rosette
	Aspergillus flavus
	aphids
Humid and sub-humid sub-region[‡]	
Maize[§]	stem borers
	virus
	larger grain borer
Cassava[§]	green mite
	grasshoppers
	weeds
Rice[§]	stem borers
	rodents
	weeds
Plantain[§]	plantain weevil
	sigatoka disease
	nematodes
	weeds
Cowpea[§]	maruca
	sucking bugs
	bruchids
	thrips

* In order of priority.
[†] Sudano-Sahelian sub-region: Burkino Faso, The Gambia, Mali, Niger, Senegal.
[‡] Humid and sub-humid sub-region: Benin, Côte d'Ivoire, Ghana, Guinea-Bissau, Nigeria, Sierra Leone, Togo, Zaire.
[§] All crops are rainfed and rice is also irrigated. Sweet potato, sugar-cane, sorghum and vegetables (not included) are priorities in some sub-regional countries.
Pesticide use is low in maize, rice and plantain and high in cowpea.
The incentive and potential for IPM in all these crops is good.

Discussion

Following the presentation of the sub-regional working group reports there was a brief discussion which focussed mainly on issues of pesticide subsidies and donations and constraints to IPM implementation related to questions of gender.

In order to achieve more rational pesticide use, the suggestion was made that international donors and national governments should be encouraged to take greater account of IPM programmes. It was also suggested that pesticide donations be sold and the money earned put towards the research and implementation of IPM initiatives.

C. Malena expressed concern that the issue of gender had not been raised during discussions about constraints to the implementation of IPM and initiatives for overcoming these constraints. She pointed out that although women are responsible for the majority of Africa's agricultural production, female researchers and extension agents are extremely rare, and some women farmers continue to have limited access to extension facilities. She noted several cases where extension activities had failed, or not been adopted, because information was provided to the male head of the household while it was the women who were responsible for the agricultural tasks in question. She suggested that for reasons of project efficiency, as well as gender equality, special efforts must be made to ensure the inclusion of women in the development and implementation of IPM.

SUMMARIES OF COUNTRY ACTION PLANS

A review was provided of the action plans produced by each country. An example of the time frame for the different activities was discussed in relation to the action plan proposed for Ghana, in which the involvement of different services and organizations was identified.

Detailed action plans for some countries are given below.

In order to summarize the different plans, a review was undertaken of the priorities identified by different countries in the first five years of a 15-year plan. This review highlighted the importance given to the need for national governments to take the necessary initiatives to promote IPM. In addition, it was widely recognized that IPM research undertaken by different institutions should be co-ordinated and the available information services improved. As seen in the summary matrix, many countries gave high priority to involving and training farmers in IPM work, and that the precise research aims for IPM studies should be clarified at an early stage.

Although several countries believe that they already have a thorough understanding of the existing crop and pest situation, others identified a need for future research into these areas. The action plans prepared by each country were to have served as a basis upon which national IPM programmes could be discussed and initiated.

INDIVIDUAL ACTION PLANS BY COUNTRY

The Gambia

Year 1. Education of policy-makers and collaborators.

Background   Pilot programmes, individual discussions about IPM packages etc.

Objective   To convince policy-makers that IPM strategies are both workable and beneficial.

Strategy   Field demonstrations, field days etc.

Institutions   Research and extension institutions, NGOs, women farmers' organizations etc.

Target   Policy-makers, influential individuals, women's groups, international organizations etc.

Years 1–5. Study the need for IPM and develop a plan at a national and sub-regional level.

Background   Given the cost, unavailability of pesticides and dangers of pesticide use there is a need for IPM.

Objective   Catalogue national IPM requirements and develop a working plan.

Strategy   Survey existing trends, identify constraints and devise solutions.

Institutions   Ministry of Agriculture, IITA, INSAH, DFPV, Gambia College, NARB, NGOs and women's organizations.

Target   Government, NGOs, IARCs and women's organizations.

Years 2–10 and 12–15. Strengthen national capabilities for IPM.

Background   Maintenance, repair and replacement of existing equipment inherited from terminated CILSS IPM project.

Objective   Provision and replacement of laboratory equipment and apparatus.

Strategy   Seek donor assistance to supplement local funds to buy equipment.

Institutions   Government, international donor community, NGOs resident in the country, women's organizations and projects within the country.

Target   International donor community.

Years 1–15. Training at all levels (researchers, technicians, male and female farmers).

Background   There is presently an acute manpower shortage at all levels to conduct IPM activities.

Objectives	Training researchers, senior staff, middle-level personnel, field agents and farmers in order to strengthen scientific and technical forces.
Strategy	Training of new research staff, refresher courses for existing staff, similar courses for middle-level personnel and field agents as well as basic technical courses for farmers.
Institutions	Higher institutes of learning, Gambia College, DFPV, WARDA, IITA, ICRISAT, NGOs and departments of research and extension.

Ghana

Objective	To develop and disseminate effective and sustainable IPM packages that will reduce the effects of pests and diseases and pesticide usage for the maintenance of a clean environment.
Justification	As a tropical country, the environmental conditions in Ghana are favourable for pests and diseases, and crop production is affected. In an attempt to control these pests farmers use many pesticides to the detriment of their own and the consumers' health, with serious damage to the environment. This reinforces the need to develop alternative pest management strategies for efficient and sustainable crop protection.
Progress achieved to date	There has been little progress achieved to date. Expertise is available at the sub-regional level from IARCs and NGOs, and locally from NARCs, universities, ministries, Global 2000 etc.

Priorities

Phase 1

(a) Develop a well-defined national policy on agricultural development with emphasis on IPM.

(b) Central co-ordination of IPM research, development and implementation;
  • institutional responsibilities must be clearly defined
  • evolution of institutional dialogue between NARCs and IARCs in IPM
  • adequate financial support for IPM with possible donor involvement.

(c) Pest surveys and identification.

(d) IPM packages to take into account socio-cultural constraints and potentials of farmers, hence the need for socio-economic studies during the development of IPM packages.

(e) Active farmer participation in the research and implementation of IPM.

(f) Human resource development at all levels to include an increase in research experience through a mentor system.

(g) Improve information services with better access to printed materials and an African journal of IPM.

(h) Eco-system analysis.

(i) Development and implementation of interventions.

(j) Recruitment of additional staff where needed.

Phases 2 and 3

(a) Continue development of IPM packages taking account of the socio-cultural constraints and potentials of farmers by including socio-economic studies.

(b) Train farmers and extension workers to take an analytical approach to problem management and to appreciate IPM activities better.

(c) Introduce a reward system for productive scientists.

(d) Technicians to ensure co-ordination and sustainability of all agricultural development policies.

(e) Continue to encourage active farmer participation in the initiation of research and implementation of IPM.

(f) Continue staff development and increase research experience through the mentor system.

Guinea-Bissau

Years 1–5. Develop and strengthen national capabilities for IPM.

Guinea-Bissau currently has no institutions or organizations involved in IPM, so national capacities for IPM will be developed and strengthened through collaboration with sub-regional, regional and international institutions such as INSAH, WARDA, ICRISAT and IITA.

Years 1–11. Identification of IPM needs and involvement with emphasis on the participation of women.

IPM programmes for the following pests and crops will be developed:

- cultural control of *Striga* on sorghum-millet-maize-based cropping systems (PASCON)
- control of storage pests using natural plant insecticides
- cultural control of scarabaeid beetle *Heteronychus oryzae* on rain-fed rice (WARDA)
- biological control of the stem borer *Plutella xylostella* through the importation of natural enemies from Cape Verde
- restricted use of insecticides to microbiological insecticides on vegetables
- biological control of cassava green mite with predators from IITA
- biological control of mango mealybug in collaboration with IITA

These pests are important because field losses on these crops are high and the crops are staple foods.

Years 1–12. Farmer training and involvement with emphasis on the participation of women.

Most IPM work will be carried out in farmers' fields, so their involvement is very important. Extension agents will receive training in IPM concepts to help improve farmer awareness and to ensure the farmers' active participation; training will concentrate on practical aspects using posters, audio-visual aids and field demonstrations. Women's participation in training will be encouraged, especially in crops like rice and vegetables. Collaboration with NGOs, both inside and outside the country, will help with training and dissemination of information.

Years 8–15. Effective co-ordination of research on IPM at the national level.

After five years of IPM implementation, there will be a need for co-ordination of IPM activities between research stations, plant protection services, NGOs and farmers (both male and female). There will be a co-ordination committee which will meet regularly to discuss research proposals, achievements and implementation in farmers' fields. The co-ordination committee should be informed on regional and sub-regional activities and achievements, and be the link between these bodies and the farmers.

Nigeria

Existing national priorities:

- control of cassava pests is already a national priority
- termites have become a nationally important pest in the southern states of the country; IPM promises to be the most important control strategy for this pest.

Nigeria lacks both central co-ordination of IPM strategies and initiatives for their implementation. The following 15-year plan is proposed to work towards the successful adoption of IPM in Nigeria.

Years 1–3

Establishment of central co-ordination for IPM activities with the appointment of a co-ordinator in the Federal Department of Agriculture who will liaise with all those involved in developing and implementing IPM strategies.

Years 3–5

(a) Surveys to quantify the extent and degree of damage nationwide.

(b) Identification of the important pests using technical expertise already available in Nigeria.

Years 6–8

Experimental investigation and development of IPM strategies for termites in cassava using combinations of chemical treatment, resistant materials, good field sanitation, early planting and weeding etc.

Specific initiatives for the following crop/cropping systems

(a) Control of *Striga* in sorghum-millet-maize-based cropping systems

This is the predominant mixed cropping system in the country and *Striga* is the most important biotic constraint. The present situation is as follows:
- tolerant maize varieties are available
- tolerant sorghum varieties for the Sudano-Sahelian zone are available
- fertilizer rates can be adjusted to enhance the tolerance of maize, sorghum and millet
- post-emergence herbicide technology is available
- no pre-emergence herbicides are available
- few tolerant materials for millet
- no resistant varieties for any of the crops
- no tolerance for sorghum in Guinea Savanna
- research-managed on-farm testing in progress

Proposals

- To fill existing gaps in research and develop complete IPM packages through collaborative and well-co-ordinated projects at international, national and regional levels.

- To conduct on-farm testing of appropriate technologies, including chemical and cultural techniques.

- To train farmers and extension workers in IPM for *Striga*, and set up research-extension farmer links, to implement IPM in the field.

(b) Control of *Striga* and *Alectra* in cowpea

Cowpea is the most important pulse crop in Nigeria, and parasitic weeds are one of the most important constraints to cowpea production. Cowpea is grown in cereal dominated mixed cropping systems, and also alone. The present situation is as follows:

- there is no commercial variety with resistance to either of these parasitic weeds

- sources of resistance to both *Striga* and *Alectra* have been identified

- resistance has been incorporated into a few locally adapted varieties

- on-farm testing of resistant varieties began in 1992.

Proposals

- To fill gaps in research through collaborative projects with IITA, SAFGRAD and other relevant Nigerian research institutions and work towards developing complete IPM packages.

- To conduct on-farm testing of appropriate IPM technology, including biological and cultural methods of control.

- To train both male and female farmers and extension workers in *Striga* and *Alectra* IPM.

(c) IPM of groundnut rosette virus disease and the aphid vector complex in groundnut

Groundnut is an important food crop and a major source of oil and protein in Nigeria. Rosette virus disease and the associated aphid complex is the most serious biological constraint to crop growth and has led to a dramatic decline in yield. At present:

- only long-season resistant varieties are available

- there is no resistance in medium- and short-season varieties, both of which are vital for the main production belt

- cultural methods of control are available

- chemical control of aphid vectors in outbreak years has not been properly co-ordinated and monitored

- current chemical control is inadequate.

Proposals

- To develop collaborative research with ICRISAT and relevant Nigerian research institutes to identify and incorporate resistance in medium- and short-duration varieties.

- To fill gaps in research towards the development of low-input IPM packages incorporating cultural and reduced chemical methods of control.

- To conduct on-farm testing of IPM packages.

- To train male and female farmers and extension agents on the IPM package, and establish well-co-ordinated research-extension-farmer links.

National IPM working group in Nigeria

There is currently no IPM Working Group in the country. Expertise to formulate IPM packages is available but scattered throughout the country in the archives of various scientists and institutions. Several crop/commodity-based working groups exist in the country, together with many active crop protection societies such as the Nigerian Society for Plant Protection, the Entomological Society of Nigeria and the Weed Science Society of Nigeria. The Federal Department of Pest Control Services is responsible for controlling pest outbreaks, and the Plant Quarantine Service is a separate department. The activities of all these groups are not co-ordinated.

Proposals

- The IPM Working Group will alert the Federal and State governments to the benefits of IPM, and explain the need for directives to the Pest Control Services to adopt an IPM approach.

- To make the public sector, agro-based enterprises and the public aware of the potential and benefit of IPM.

- To encourage the establishment and strengthening of sustainable pest control strategies nationwide.

- To provide a central database of existing pest control techniques in the country with a view to evolving IPM packages for the various crop/pest priorities.

- To fill gaps in research in the present technologies for pest control in various crops.

- To provide short-term IPM training for scientists, technicians, extension agents and farmers.

- To encourage effective links between research workers, extension specialists, farmers and policy-makers in IPM activities.

This will be achieved by:

- setting up an *ad hoc* steering committee to elaborate proposals for government consideration

- holding consultation meetings with governments to request approval and funds for a national IPM Working Group

- organizing a national workshop on IPM

- informing the Federal and State Governments, public sector enterprises and the public about the benefits of IPM

- developing action plans for IPM packages of various crops
- organizing periodic meetings for evaluation of programmes and information exchange
- publishing IPM awareness bulletins
- initiating and maintaining an IPM data bank.

The international community of research institutes and donors, national institutes and universities, government departments and private organizations will all be involved in collaborating to develop and implement strategies for IPM.

Increased awareness of the benefits of IPM should result in the more judicious use of pesticides with a reduction in the health hazards associated with their use. Another benefit of co-ordinated IPM activities nationwide will be increased food production.

SUB-REGIONAL NETWORKING

The scope and details of networking requirements were discussed by separate working groups formed for the Sudano-Sahelian and the humid and sub-humid sub-regions.

Although the approaches adopted by the two groups differed, it was agreed that the constraints to IPM implementation were common and provided a basic need for a network that will integrate the experiences of each member and pool information for the region. The network will include national support services, private institutions, NGOs and international institutions. This should improve awareness of IPM at the sub-regional level, help to identify research and extension needs and ensure that work is not duplicated unnecessarily.

Once common problems are identified, networking could be achieved through regular meetings, the establishment of a steering committee and a newsletter. It was initially proposed that two networks be established within West Africa, one based in the Sudano-Sahelian sub-region and the other in the humid and sub-humid sub-region. A discussion ensued about the existing network organized by CILSS in Sahelian countries.

In conclusion it was agreed that any new network should be based upon identified national priorities and designed to complement and reinforce existing networking facilities. Periodic evaluations should be included in the networking plan to provide a guarantee of credibility and efficiency for members and donors.

Recommendations

POLICY

(a) Government commitment to IPM is essential for its development and implementation. In order to promote awareness within government circles, seminars for decision-makers on the necessity of IPM development and implementation should be organized by regional organizations with assistance from the donor community.

(b) Emphasis should be placed by donors and recipient countries on developing and supporting policies and practices which will help to promote and implement IPM.

(c) In view of the serious constraints on IPM development and implementation caused by the free distribution of pesticides, governments should be encouraged to introduce a system in which farmers contribute financially to the cost of pesticides and their application.

(d) In the case of pest emergencies, improved and standardized procurement and supply procedures for pesticides must be adhered to, e.g. the draft *FAO Guidelines on the Procurement and Tender of Pesticides*, to minimize the delivery of pesticides in excess of the need. In addition, a system of accountability on the distribution and use of pesticides should be introduced.

DEVELOPMENT AND IMPLEMENTATION OF IPM

(a) IPM is an essential contribution towards attaining sustainable agriculture. It is a component of integrated crop production and should be developed and implemented within the context of agricultural production systems. More attention should be given to the growing of healthy crops as the basis of IPM.

(b) IPM programmes should be decided with farmer involvement based on studies of their socio-economic environments with special emphasis on gender issues.

(c) In addition to identifying objectives and strategies, programmes for the development and implementation of IPM should also include mechanisms for monitoring progress. The successful development of IPM requires a high level of flexibility, and methods of monitoring progress should be modified as required and include action thresholds, and cost/benefit assessments.

(d) Governments should be encouraged to involve NGOs working with national research and extension agencies in the promotion and implementation of IPM.

INSTITUTIONAL CAPACITY BUILDING

(a) Governments should be encouraged to prepare medium- and long-term strategic plans to strengthen plant protection programmes. These will cover research, training, extension, implementation, regulation and pest control. Such plans are necessary to set priorities for national and donor support.

(b) IPM should be institutionalized through the creation of regional IPM networks as and when required, and existing networks should be strengthened.

Summary of country action plans

		Guinea Bissau	Sierra Leone	Zaire	Togo/ Benin	Nigeria	Ghana
Political	National policy on IPM	+		+		+	
	Funding for IPM			+	+		
	IPM must be continued/sustained		+				
	Payment of counterparts by donors				+		
Institutional	Clearly defined institutional responsibilities		+	+			
	Centrally co-ordinate IPM resource work		+			+	+
	Better dialogue between departments involved in IPM		+				
	Involve private sector			+			
Socio-economics	Re-inforce extension service				+		
	Involve and train farmers		+	+		+	+
	Transfer government subsidies on pesticides			+		+	
	IPM must be cost-effective				+		
Technical	More research into pests and eco-systems			+		+	+
	Identify research needs for IPM		+		+		
	Improve information services			+	+	+	+
	New equipment					+	
	Training			+		+	

PARTICIPANTS

Benin
Chakirou LAWANI
Entomologist
Service Protection de Végétaux
Direction de l'Agriculture
B. P. 58
Porto Novo

Tel 229 21 44 13
Fax 229 21 39 37

André KATARY
Directeur de la Recherche Coton et Fibres
Direction de la Recherche Agricole
B. P. 715
Cotonou

Tel 229 31 34 46
Fax 229 30 05 91

Burkino Faso
Yacouba SERE
Plant Pathologist – Head of Rice Program
INERA – B.P. 910
Bobo Dioulasso

Tel 226 98 23 29
Fax 226 98 20 42

Cote d'Ivoire
Nanga COULIBALY
Directeur Dept Cafe Cacao – IDEFOR
01 B.P. 1827
Abidjan 01

Tel 225 30 30 32
Fax 225 22 69 85

The Gambia
Dodou C.A. JAGNE
Department of Agriculture Services
Cape St. Mary

Tel 220 95 312
Fax 220 95 413
Telex 2312 MINAGRIC

Sainey KEITA
Department of Agricultural Research
Yundum

Tel 220 82 875
Fax 220 82 875

Ghana
J.F. ABU
Deputy Secretary (Western Region)
Ministry of Agriculture
P.O. Box 304
Sekondi

G.A. DIXON
Head, Plant Protection and Regulatory Services
Ministry of Agriculture
P.O. Box M37
Accra

Kwame AFREH-NUAMAH
Pesticide Application Technologist
Faculty of Agriculture
Agricultural Research Station, Kade
University of Ghana, Legon
Accra

Telex 2556 UGGH

Charles Y. BREMPONG-YEBOAH
Lecturer
Crop Science Department
University of Ghana, Legon
Accra

Telex 2556 UGGH

124

Haruna BRAIMAH
Crops Research Institute
P.O. Box 3785
Kumasi

A.R. CUDJOE
Plant Protection and Regulatory Service
Ministry of Agriculture
P.O. Box M37
Accra

S. KORANG-AMOAKAH
Director, Department of Agriculture and Extension Services
P.O. Box M37
Accra

F. OFORI
Director, Department of Crops
Ministry of Agriculture
P.O. Box M37
Accra

R. Nana Yaw OBENG
Managing Director
AGRI-MAT Ltd
P.O. Box 15097
Accra-North

L.O. OPARE
Dizengoff Ghana Ltd
Accra

Eunice ADAMS
Senior Agricultural Officer
Plant Protection and Regulatory Service
Ministry of Agriculture
P.O. Box M37
Accra

Ruth OSEI-BONSU
Assistant Agricultural Officer
Plant Protection and Regulatory Service
Ministry of Agriculture
P.O. Box M37
Accra

Guinea-Bissau
Mustafa CASSAMA   Tel 245 21 10 41
Crop Protection Director   Fax 245 22 10 71
MDRA/DSPV
C.P. 71
Bissau

Lourenço ABREU   Tel 245 21 10 41
Entomologist, MDRA – DSPV   Fax 245 22 10 71
P.O. Box 71
Bissau

Niger
Hamidou LAZOUMAR   Tel 227 74 11 57
Direction Protection des Végétaux   Fax 227 73 30 20
B.P. 323
Niamey

Maiquizo MOUNKAILA   Tel 227 44 10 08
Chef du Centre de Lutte Biologique
INRAN
B.P. 123
Agadez
or
B.P. 429   Tel 227 72 27 14/19
Niamey   Telex 5201 NI

Nigeria
S.M. MISARI   Tel 234 69 50571 – 4PBX
Deputy Director   Tel 234 69 50681 – 4PBX
Institute for Agricultural Research   Telex 75242 AGFCZX
Ahmadu Bello University   or 75248 NITEZNG
Samaru PMB 1044
Zaria

T.N.C. ECHENDU   Tel 234 88 22 01 88
Co-ordinator National Biological Control Program
c/o National Root Crops Research Institute,
Umudike – Umuahia

Alpho Mgbanu EMECHEBE   Tel (069) 50571
Professor/Dean of Faculty   Telegram AGRISEARCH,
Department of Crop Protection   ZARIA, NIGER
Institute of Agricultural Research/Faculty of Agriculture   Telex 75248 NITEZNG
Ahmadu Bello University   or 75242 AGRCZX
Samaru, Zaria

Segun T.O. LAGOKE
Professor/Co-ordinator
Pan Africa Striga Control Network
Department of Agronomy, Institute for Agricultural Research
Ahmadu Bello University
Samaru, PMB 1044
Zaria

Senegal
Moctar WADE   Tel 221 73 60 50
ISRA-CNRA-BAMBEY   Fax 221 73 60 52
B.P. 53
Diourbel

Sierra Leone
C.B. SESAY   Tel 232 41 525
Director-General of Agriculture,   Telex 3418 PEMSU
Forestry and Fisheries
c/o Ministry of Agriculture
Freetown

B.D. JAMES   Tel 232 41 755
Co-ordinator, National Biological Control Programme   Fax 232 22 47 13
c/o FAO
P.O. Box 71
Freetown

Julia ROBERTS   Tel c/o Cecil Farmer
Department of Crop Protection   SLET 232 22 41 17
Institute of Agricultural Research   Fax   232 22 44 39
P.O. Box 540, Freetown

Togo

Dovi AGOUNKE
Directeur, DPV
B.P. 1263
Cacaveli, Lomé

Tel   228 21 37 73
Fax 228 21 10 08

Zaire

Nsiama SHE H.D.
Director Cassava Programme
Ministère de l'Agriculture
B.P. 11635
Kinshasa

Tel   243 12 27 242
Telex 21536 LASCO ZR

AFRICARE

Daniel Eric GERBER
B.P. 1792
Bamako
Mali

Tel   223 22 37 03
Fax 223 22 02 91

APEMAF

Alfred LATIGO
Director-General
P.O. Box 14126
Nairobi
Kenya

Tel   254 2 44 70 78
Fax 254 2 44 60 80

CARE

Darius MIDDAH
ANR Co-ordinator
B.P. 143
Maradi, Niger

Tel   227 41 07 05
Fax 227 41 07 30

J-M. VIGREUX
ANR Program Co-ordinator
B.P. 143
Maradi, Niger

Tel   227 41 07 05
Fax 227 41 07 30

CILSS

Daoule Diallo BA
Coordonnatrice UCTR/PV
B.P. 1530
Bamako

Tel   223 22 46 81
Fax 223 22 59 80

Sankung B. SAGNIA
Entomologist
B.P. 12625
Niamey
Niger

Tel   227 73 21 81
Fax 227 73 21 81

FAO

Gerard G.M. SCHULTEN
Senior Officer (Entomology)
Plant Protection Service
Via delle Terme di Caracalla
Rome, Italy

Tel   39 6 59 97 35 51
Fax 39 6 57 97 31 52

Sulayman S. M'BOOB
Senior Crop Protection Officer
FAO Regional Office for Africa
P.O. Box 1628
Accra, Ghana

Tel   233 21 66 68 51 to 54
Fax 233 21 66 84 27

Rik G.J.K. HOEVERS
Weed Management
FAO Regional Office for Africa
P.O. Box 1628
Accra, Ghana

Tel   233 21 66 68 51 to 54
Fax 233 21 66 84 27

GRAT
Youssouf SANOGO
B.P. 2502
Bamako
Mali

Tel   223 22 43 41
Fax 223 22 23 59
(attention GRAT)

GTZ
Jean Charles HEYD
GTZ/Benin
B.P. 7021
Cotonou
Benin

Tel   229 21 32 93
Fax 229 30 13 65

Horst U. FISCHER
GTZ/Togo
B.P. 1263
Lomé

Tel   228 21 37 73
Fax 228 21 10 08

ICRISAT
Ousmane YOUM
Principal Cereals Entomologist
ICRISAT Sahelian Center
B.P. 12404
Niamey
Niger

Tel   227 72 25 29
Fax 227 73 43 29

IDS
Carmen Celeste MALENA
Institute of Development Studies
University of Sussex
MPI4 Falmer, Brighton
United Kingdom

Tel   44 273 606 261

IIBC
Chris LOMER
IIBC Representative
IITA, B.P. 08–0932
Cotonou
Benin

Tel   229 3601 88
Fax 229 30 14 66

IITA
S. YANINEK (acarologist)
B.P. 08–0932
Cotonou
Benin

Tel   229 36 01 88
Fax 229 30 14 60

Manuele TAMO
Entomologist
B.P. 08–0932
Cotonou
Benin

Tel   229 36 01 88
Fax 229 30 14 66

Winfred HAMMOND
B.P. 08–0932
Cotonou
Benin

Tel   229 36 01 88
Fax 229 30 14 66

Kitty F. CARDWELL
PMB 5320
Ibadan
Nigeria

Fritz SCHULTESS   Tel 229 36 01 88
Scientist, IITA   Fax 229 30 14 66
Cotonou
Benin

Petra PARNITZKI   Tel 229 36 01 88
GTZ/IITA   Fax 229 30 14 66
B.P. 08–0932
Cotonou
Benin

Rodale Senegal
Amadou Makhtar DIOP   Tel 221 511028
Director   Fax 221 511670
B.P. A237   CGNET GCD0013
Thies, Senegal

USAID
Amy J. DREEVES   Tel 223 22 36 02
USAID – DAI/OHV Mali   Fax 223 22 39 33
B.P. 34
Bamako
Mali
or
Oregon State University   Tel 303 493 17 76
Dept of Entomology   Fax 503 737 30 80
2046 Cordley Hall
Corvallis, OREGON 97331–2907
USA

Kondo Mahaman SANI   Tel 227 73 35 08
AELGA Project Co-ordinator Assistant   Fax 227 72 39 18
USAID/Niger
c/o American Embassy
B.P. 11201
Niamey
Niger

US Peace Corps
Kenneth BYRD   Tel 202 606 34 02
Agriculture Program Specialist   Fax 202 606 30 24
1990 K St, NW
Washington D.C. 20524,
USA

WARDA
Anthony YOUDEOWEI   Tel 225 63 45 14
Director of Training and Communications   Fax 225 63 47 14
01 B.P. 2551   22 78 65
Bouaké
Côte d'Ivoire

Abdoul Aziz SY   Tel 225 63 45 14
Senior Pathologist   Fax 225 63 47 14
01 B.P. 2551
Bouaké
Côte d'Ivoire

IPM Working Group
Patricia C. MATTESON
IPM Consultant/Facilitator
Department of Entomology
Iowa State University
Ames, IA 50011 – 3140, USA

Malcolm ILES
Secretary, IPM Working Group
Natural Resources Institute, Chatham Maritime, Kent
ME4 4TB, United Kingdom

Tel 634 88 00 88
Fax 634 88 30 54

Francois VICARIOT
ORSTOM
Delegue aux Affaires Internationales
213 rue Lafayette
75480 Paris cedex 10, France

Tel 331 48 03 77 20
Fax 331 40 36 23 85

United Kingdom
Thomas WOOD
Natural Resources Institute
Chatham Maritime, Kent
ME4 4TB

Tel 634 88 00 88
Fax 634 88 00 66

Christopher WEST
ODA Advisor
Victoria Street
London

Tel 71 91 77 000

USA
Walter I. KNAUSENBERGER
Environmental Analyst
Department State, USAID
AFR/ARTS/FARA
Washington D.C. 20523 – 1515, USA

Tel 703 235 38 26
Fax 703 235 38 05

Robert C. HEDLUND
406 L SA-18
USAID/R&D/AGR
Washington D.C. 20523 – 1809, USA

Tel 703 875 4233
Fax 703 875 5344

SUB-REGIONAL WORKING GROUPS

Group I
Sudano-Sahelian Working Group – Members

Co-ordinator: Dr Anthony Youdeowei, WARDA
Rapporteur: S. Sagnia, DFPV, Niamey, Niger
              Amadou Diop, RODALE, Senegal

Name	Country	Organization
Youssouf Sanogo	Mali	GRAT
Daniel Gerber	Mali	AFRICARE
Amy Dreeves	Mali	DAI-USAID/OHV
Maïquizo Mounkaila	Niger	CLB/INRAN – Niger
Kondo Mahaman Sani	Niger	USAID
Darius Middah	Niger	CARE International
J.M. Vigreux	Niger	CARE International
Segun Lagoke	Nigeria	IAR/ABU/PASCON
S.M. Misari	Nigeria	IAR/ABU Zaria
Alpho Emechebe	Nigeria	IAR/ABU
Chris Lomer	Benin	IIBC
Kitty Cardwell	Benin/Nigeria	IITA
Sainey Keïta	The Gambia	DAR
Dodou Jagne	The Gambia	Department of Agricultural Services
Rik Hoevers	Ghana	FAO
Mustapha Cassama	Guinea-Bissau	C.P. Services
Lourenco Abreu	Guinea-Bissau	C.P. Services
Moctar Wade	Senegal	ISRA-CNRA-BAMBEY

Group II

Name	Country	Organization
Chakirou Lawani	Benin	SPV
André Katary	Benin	DRCF/DRA
Yacouba Sere	Burkina Faso	INERA
Nanga Coulibaly	Côte d'Ivoire	IDEFOR
J.F. Abu	Ghana	Ministry of Agriculture
G.A. Dixon	Ghana	Ministry of Agriculture
Kwame Afreh-Nuamah	Ghana	ARS/University of Ghana
Charles Brempong-Yeboah	Ghana	University of Ghana
Haruna Braimah	Ghana	CRI
A.R. Cudjoe	Ghana	Ministry of Agriculture
S. Korang-Amoakah	Ghana	DAES
F. Ofori	Ghana	Ministry of Agriculture
R. Nana Yaw Obeng	Ghana	AGRI-MAT
L.O. Opare	Ghana	Dizengoff
Eunice Adams	Ghana	Ministry of Agriculture
Ruth Osei-Bonsu	Ghana	Ministry of Agriculture
Hamidou Lazoumar	Niger	DPV
T.N.C. Echendu	Nigeria	NRCRI
C.B. Sesay	Sierra Leone	AFF/Ministry of Agriculture
B.D. James	Sierra Leone	NBCP/FAO
Julia Roberts	Sierra Leone	IAR
Dovi Agounke	Togo	DPV
Nsiama She H.D.	Zaire	Ministry of Agriculture
Alfred Latigo	Kenya	APEMAF
Daoule Diallo BA	Mali	CILSS
Sankung B. Sagnia	Niger	CILSS

Name	Country	Organization
Jean Charles Heyd	Benin	GTZ
Horst Fischer	Togo	GTZ
Ousmane Youm	Niger	ICRISAT
S. Yaninek	Benin	IITA
Manuele Tamo	Benin	IITA
Winfred Hammond	Benin	IITA
Fritz Schultess	Benin	IITA
Petra Parnitzki	Benin	IITA
Amadou Makhtar Diop	Senegal	Rodale Senegal
Kondo Mahaman Sani	Niger	Rodale Senegal
Kenneth Byrd	USA	US Peace Corps
Anthony Youdeowei	Côte d'Ivoire	WARDA
Abdoul Aziz Sy	Côte d'Ivoire	WARDA

Group III
'Floating' Resource Persons

Name	Country	Organization
Malcolm Iles	United Kingdom	NRI
Thomas Wood	United Kingdom	NRI
Chris West	United Kingdom	ODA
Robert Hedlund	USA	USAID
Walter Knausenberger	USA	USAID
Patricia Matteson	USA	Iowa State University
Francois Vicariot	France	ORSTOM
Carmen Malena	United Kingdom	IDS
Gerard Schulten	Italy	FAO
Sulayman M'Boob	Ghana	FAO

ABBREVIATIONS USED IN THIS REPORT

APEMAF	African Pest and Environment Management Foundation
AVRDC	Asian Vegetable Research and Development Centre
CGIAR	Consultative Group for International Agricultural Research
CIAT	International Centre for Tropical Agriculture
CIDA	Canadian International Development Agency
CILSS	Comité Permanent Inter-Etats de Lutte contre la Secheresse dans le Sahel
CIMMYT	International Maize and Wheat Improvement Centre
CIRAD	Centre de Coopération International en Recherche Agronomique pour la Developpement
CPS	Crop Protection Service
CRAR	Centre de Régénération des Resources Agricoles
CRSP	Collaborative Research Support Programme
CTA	Technical Centre for Agricultural and Rural Co-operation
CTC	Certificate Training Centre
DAAD	German Academic Exchange Services
DFPV	Département de Formation en Protection des Végétaux
DGIS	Directorate General for International Co-operation
DHV	Development of the Haute Valley
DLCO-EA	Desert Locust Control Organization for Eastern Africa
DPV	Département de Protection Végétaux
DPVC	Department of Plant Protection and Packaging
EC	European Community
FAO	Food and Agriculture Organization of the United Nations
FDPCS	Federal Department of Pest Control Services
GRAT	Groups de Recherche et d'Applications Techniques
GTZ	Deutsche Gesellshaft fur Technische Zusammenarbeit
IAR	Institute of Agricultural Research
IARC	International Agricultural Research Centre
ICIPE	International Centre for Insect Physiology and Ecology
ICRISAT	International Crops Research Institute for the Semi-Arid Tropics
IDRC	International Development Research Centre
IER	Institut d'Economie Rurale
IIBC	International Institute of Biological Control
IITA	International Institute of Tropical Agriculture
INRA	Institut National de Recherche Agronomique
INRAN	National Institute for Agricultural Research in Niger
IPM	integrated pest management
IPMWG	Integrated Pest Management Working Group

IRRI	International Rise Research Institute
ISRA	Institut Sénégalese de Recherches Agronomiques
NARES	National Agricultural Research and Extension Systems
NARP	National Agricultural Research Project
NARS	National Agricultural Research Systems
NGO	non-governmental organization
NIHORT	National Horticultural Research Institute
NRI	Natural Resources Institute
OAU	Organization of African Unity
OCLALAV	Organisation Commune de Lutte Antiacridienne et de Lutte Antiaviare
ODA	Overseas Development Administration
OHV	Operation Haute Vallee
ORSTOM	Office de la Recherche Scientifique et Technique Outre-Mer
PASCON	Pan-Africa Striga Control Network
PCV	Peace Corps Volunteer
RRCA	Regional Research Co-ordination Agency
SADCC	Southern African Development Co-ordination Conference
SNPV	Service National de Protection des Vegetaux
SNV	Service National de Vulgarisation Agricole
UCTR/PV	Unité de Coordination Technique Regionale en Protection Végétaux
UNDP	United Nations Development Programme
UNEP	United Nations Environment Programme
UNICEF	United Nations Children's Fund
USAID	United States Agency for International Development
WARDA	West African Rice Development Association
WHO	World Health Organization

Printed by Hobbs the Printers of Southampton